Studies in Systems, Decision and Control

Volume 446

Series Editor

Janusz Kacprzyk, Systems Research Institute, Polish Academy of Sciences, Warsaw, Poland

The series "Studies in Systems, Decision and Control" (SSDC) covers both new developments and advances, as well as the state of the art, in the various areas of broadly perceived systems, decision making and control–quickly, up to date and with a high quality. The intent is to cover the theory, applications, and perspectives on the state of the art and future developments relevant to systems, decision making, control, complex processes and related areas, as embedded in the fields of engineering, computer science, physics, economics, social and life sciences, as well as the paradigms and methodologies behind them. The series contains monographs, textbooks, lecture notes and edited volumes in systems, decision making and control spanning the areas of Cyber-Physical Systems, Autonomous Systems, Sensor Networks, Control Systems, Energy Systems, Automotive Systems, Biological Systems, Vehicular Networking and Connected Vehicles, Aerospace Systems, Automation, Manufacturing, Smart Grids, Nonlinear Systems, Power Systems, Robotics, Social Systems, Economic Systems and other. Of particular value to both the contributors and the readership are the short publication timeframe and the world-wide distribution and exposure which enable both a wide and rapid dissemination of research output.

Indexed by SCOPUS, DBLP, WTI Frankfurt eG, zbMATH, SCImago.

All books published in the series are submitted for consideration in Web of Science.

Sarah El Himer · Mariyam Ouaissa ·
Abdulrahman A. A. Emhemed · Mariya Ouaissa ·
Zakaria Boulouard
Editors

Artificial Intelligence of Things for Smart Green Energy Management

Springer

Editors
Sarah El Himer
Department of Electrical Engineering
Sidi Mohamed Ben Abdellah University
Fez, Morocco

Abdulrahman A. A. Emhemed
Department of Electrical Engineering
Bright Star University
El-Brega, Libya

Zakaria Boulouard
University of Hassan II Casablanca
El Mansouria, Morocco

Mariyam Ouaissa
Moulay Ismail University
Meknes, Morocco

Mariya Ouaissa
Moulay Ismail University
Meknes, Morocco

ISSN 2198-4182 ISSN 2198-4190 (electronic)
Studies in Systems, Decision and Control
ISBN 978-3-031-04850-0 ISBN 978-3-031-04851-7 (eBook)
https://doi.org/10.1007/978-3-031-04851-7

This Springer imprint is published by the registered company Springer Nature Switzerland AG
The registered company address is: Gewerbestrasse 11, 6330 Cham, Switzerland

Preface

Energy is considered a very important ingredient in all human activities such as health, food production, education, industrial production, and transportation. Modern energy services are a powerful engine of economic and social development. In fact, no country has ever succeeded in developing well beyond a subsistence economy without ensuring a minimal access to energy services for a large part of its population. Green energy is defined as a primary energy source that produces a relatively small amount of pollutants when transformed into final energy and then used as such. Green energy is obtained from renewable sources such as sunlight, wind, biomass, and geothermal energy. They neither affect general health, nor pollute the environment, nor represent a burden on the environment. For this reason, green energy is the technology that will ensure a clean and healthy environment to the future generations.

For saving energy in buildings, vehicles, communication networks, and industries, it is important to develop efficient Energy Management Systems (EMS). Like energy production, energy saving is also equally important because the amount of energy saved can be considered as the amount of energy generated. Therefore, it is extremely important to save energy as much as possible in both consumer and industrial context scenarios.

With the advancements in technology, the green energy sector has made a significant progress in the last decade. However, there are still a few challenges in this sector that can be addressed with the help of emerging technologies. Technologies like Artificial Intelligence (AI) and Machine Learning (ML) can analyze the past, optimize the present, and predict the future. AI in the renewable and green energy sector can resolve most of the challenges these sectors are facing.

In order to produce low-cost, smart, and efficient solutions, advanced green energy management systems must use cutting-edge technologies such as artificial intelligence and the Internet of Things. On one hand, Artificial Intelligence helps the system to optimize power consumption, resulting in a more energy-efficient system; on the other hand, the Internet of Things allows the system to be wire-free and operate with more flexibility.

Intelligent solar energy systems, intelligent wind energy systems, and energy efficiency oriented research in intelligent electric cars are just a few of the subjects covered in this book. Artificial intelligence technologies such as artificial neural networks, genetic algorithms, fuzzy logic, and a mix of the aforementioned in hybrid systems are used in all of these disciplines and more. Energy management, equipment monitoring, security, fire management, life safety, and factory automation are some of the application options.

This book is intended to assist in the development of low-cost, smart, and efficient green energy solutions. We will also examine the problems that various nations have in these areas and provide solutions based on AI and IoT.

The first chapter of this book starts with an introduction to Artificial Intelligence and the Internet of Things and their role in renewable energy. The authors in chapter "Design of a Switching Module for Electricity Supply from Solar to Grid by Sensing Light Intensity" suggest a new design of a switching module for electricity supply from solar to grid by sensing light intensity, while in chapter "Nonlinear Backstepping Control of a Grid-Connected Doubly Fed Induction Generator Wind Turbine", introduce a new nonlinear backstepping control of a grid-connected doubly fed induction generator wind turbine. The authors of chapter "Improving LVRT Capability of a Wind Turbine During a Voltage Sag in the Electrical Network" suggest a new approach to improve the LVRT capability of a wind turbine during a voltage sag in the electrical network.

In chapter "Optimization of Two Hybrid Micro-Concentrator Photovoltaic Systems for Car-Roof Application", the authors suggest a new approach to optimize a two hybrid micro-concentrator photovoltaic systems for car roofs, while in chapter "Machine Learning-Based Maximum Power Point Tracking Technique for Concentrated PV/TEG System Under Non-uniform Environmental Conditions" they provide a machine learning-based maximum power point tracking technique for concentrated PV/TEG system under non-uniform environmental conditions. The authors in chapter "Two-Dimensional Nanomaterials for Solar Cell Technology" discuss the potential role of two-dimensional Nanomaterials for solar cell technology, while in chapter "Improving Control for an Induction Machine Using Artificial Intelligence" suggest using Artificial Intelligence to improve the control of induction machines.

The authors of chapter "Evaluation of the Recommended Algorithms in the Internet of Things" evaluate recommendation algorithms in Internet of Things, while the author of chapter "Exposing Applications of IoT in Green Computing" exposes different applications of IoT in Green Computing.

In chapter "IoT Safe Area: An Innovative Approach for a Smart Decentralized and Hybrid Decision Support Process for Industrial Warehousing", the authors suggest an innovative approach for a smart decentralized and hybrid decision support process for industrial warehousing, while in chapter "Application of Artificial Intelligence (AI) in Supply Chain: An Overview", they provide a general overview on different applications of Artificial Intelligence (AI) in Supply Chains.

The authors of chapter "Acorn Waste Valorization as Reinforcement in Polystyrene Composite: A Comparative Study" provide a comparative study over different approaches in Acorn Waste Valorization as reinforcement in Polystyrene Composite, while the authors of chapter "Design and Analysis of a Smart Anti-lock Braking System" provide a new design and analysis of a smart anti-lock braking system.

Fez, Morocco — Sarah El Himer
Meknes, Morocco — Mariyam Ouaissa
El-Brega, Libya — Abdulrahman A. A. Emhemed
Meknes, Morocco — Mariya Ouaissa
El Mansouria, Morocco — Zakaria Boulouard
March 2022

Contents

About the Editors

Dr. Sarah El Himer is a Ph.D. and Researcher Associate in Intelligent Systems, Georesources & Renewable Energies Laboratory. She is Ph.D., graduated in 2019 in renewable energies at the renewable energies and intelligent systems laboratory from the faculty of sciences and technology, university of Sidi Mohammed Ben Abdallah FEZ. She is a trainer and IT Consultant at IT Support and Consulting Center. She was working for faculty of science and technology FEZ as a Visiting Professor from 2015 to 2019. Dr. El Himer has made contributions in the fields of optical component of Concentrated photovoltaic and electrical vehicle. Her main research topics are Optical element for CPV, acceptance angle, optical efficiency micro CPV, Hybrid CPV. She has published over 15 papers (international journals, and conferences/workshops). She has served and continues to serve on executive and technical program committees and as a reviewer of numerous international conference and journals such as CPV conference, REEE2017".

Dr. Mariyam Ouaissa is currently a Professor at Institute specializing in new information and communication technologies, Researcher Associate and Consultant Trainer in Computer Science and Networks. She received her Ph.D. degree in 2019 from National Graduate School of Arts and Crafts, Meknes, Morocco and her Engineering Degree in 2013 from the National School of Applied Sciences, Khouribga, Morocco. She is a communication and networking researcher and practitioner with industry and academic experience. Dr. Ouaissa's research is multidisciplinary that focuses on Internet of Things, M2M, WSN, vehicular communications and cellular networks, security networks, congestion overload problem and the resource allocation management and access control. She is serving as a reviewer for international journals and conferences including as IEEE Access, Wireless Communications and Mobile Computing. Since 2020, she is a member of "International Association of Engineers IAENG" and "International Association of Online Engineering", and since 2021, she is an "ACM Professional Member". She has published more than 20 research papers (this includes book chapters, peer-reviewed journal articles, and peer-reviewed conference manuscripts), 8 edited books, and 5 special issue as guest editor. She has served on Program Committees and Organizing Committees of several conferences and events and has organized many Symposiums/Workshops/Conferences as a General Chair.

Dr. Abdulrahman A. A. Emhemed is a lecturer at Bright Star University, El-Brega, Libya. He obtained B.Sc. Degree in Control Engineering from College of Electronic Technology, Bani Walid, Libya, in 2006. M.Sc. degree of Mechatronic and Robotic Engineering from University Tun Hussein Onn Malaysia (UTHM), 2011. Ph.D. degree in Electrical Engineering from University of Technology Malaysia (UTM), 2017. He has many papers published at international conferences and journals. Dr. Emhemed is registered as Professional Engineer by the International Association of ENGineers (IAENG) and member for many international institutes and engineering associations such as Institute of Electrical and Electronics Engineering (IEEE), SCIence and Engineering Institute (SCIEI), and Hong Kong Society of Mechanical Engineers (HKSME). His main area of research includes Model Predictive Control, Fuzzy Logic Control for Mechatronics systems, and Robust PID Control Techniques and Optimization.

Dr. Mariya Ouaissa is currently a Professor at Institute Specializing in New Information and Communication Technologies, Researcher Associate and practitioner with industry and academic experience. She is a Ph.D. graduated in 2019 in Computer Science and Networks, at the Laboratory of Modelisation of Mathematics and Computer Science from ENSAM-Moulay Ismail University, Meknes, Morocco. She is a Networks and Telecoms Engineer, graduated in 2013 from National School of Applied Sciences Khouribga, Morocco. She is a Co-Founder and IT Consultant at IT Support and Consulting Center. She was working for School of Technology of Meknes Morocco as a Visiting Professor from 2013 to 2021. She is member of the International Association of Engineers and International Association of Online Engineering, and since 2021, she is an "ACM Professional Member". She is Expert Reviewer with Academic Exchange Information Centre (AEIC) and Brand Ambassador with Bentham Science. She has served and continues to serve on technical program and organizer committees of several conferences and events and has organized many Symposiums/Workshops/Conferences as a General Chair also as a reviewer of numerous international journals. Dr. Ouaissa has made contributions in the fields of information security and privacy, Internet of Things security, and wireless and constrained networks security. Her main research topics are IoT, M2M, D2D, WSN, Cellular Networks, and Vehicular Networks. She has published over 20 papers (book chapters, international journals, and conferences/workshops), 8 edited books, and 5 special issue as guest editor.

Dr. Zakaria Boulouard is currently a Professor at Department of Computer Sciences at the "Faculty of Sciences and Techniques Mohammedia, Hassan II University, Casablanca, Morocco". In 2018, he joined the "Advanced Smart Systems" Research Team at the "Computer Sciences Laboratory of Mohammedia". He received his Ph.D. degree in 2018 from "Ibn Zohr University, Morocco" and his Engineering Degree in 2013 from the "National School of Applied Sciences, Khouribga,

Morocco". His research interests include Artificial Intelligence, Big Data Visualization and Analytics, Optimization and Competitive Intelligence. Since 2017, he is a member of "Draa-Tafilalet Foundation of Experts and Researchers", and since 2020, he is an "ACM Professional Member". He has served on Program Committees and Organizing Committees of several conferences and events and has organized many Symposiums/Workshops/Conferences as a General Chair. He has served and continues to serve as a reviewer of numerous international conferences and journals. He has published several research papers. This includes book chapters, peer-reviewed journal articles, peer-reviewed conference manuscripts, edited books, and special issue journals.

Artificial Intelligence of Things (AIoT) for Renewable Energies Systems

Sarah El Himer, Mariya Ouaissa, Mariyam Ouaissa, and Zakaria Boulouard

Abstract Solar panels, wind farms, batteries, and power transformers are among the several types of distributed energy resources used in today's power system. Data connections and IoT sensors are common features of DERs, and they create a lot of data. These data may be used to increase system efficiency as well as create extra values, in addition to monitoring device status. In this study, we look at the technical benefits of AI algorithms on the expanding DER data. First, we attempted to demonstrate how artificial intelligence and the internet of things may help renewable energy systems function better. Second, AIoT (Artificial Intelligence Internet of Things) is introduced as a new sector following IoT technology. AIoT (Artificial Intelligence Internet of Things) are new technologies that combine AI with IoT to provide new opportunities in the distributed energy resources (DER) industry. Finally, this article concentrates on AIoT applications in renewable energy sources such as solar and wind.

Keywords Renewable energy · Internet of things · Artificial intelligence · AIoT

S. El Himer (✉)
Sidi Mohammed Ben Abdellah, Fez, Morocco
e-mail: sarah.elhimer@usmba.ac.ma

M. Ouaissa · M. Ouaissa
Moulay Ismail University, Meknes, Morocco
e-mail: mariya.ouaissa@edu.umi.ac.ma

M. Ouaissa
e-mail: mariyam.ouaissa@edu.umi.ac.ma

Z. Boulouard
LIM, Hassan II University, Casablanca, Morocco
e-mail: zakaria.boulouard@fstm.ac.ma

S. El Himer et al. (eds.), *Artificial Intelligence of Things for Smart Green Energy Management*, Studies in Systems, Decision and Control 446,
https://doi.org/10.1007/978-3-031-04851-7_1

1 Introduction

Artificial Intelligence of Things (AIoT) stands for Artificial Intelligence + Internet of Things, which refers to a practical use of artificial intelligence with the Internet of Things. In a Distributed Energy Resource (DER) environment, the Internet of Things (IoT) links a vast number of devices and machines to the Internet; this includes all power sources, end node and transformer switches, and so on. IoT sensors create a large quantity of data that reflects the state of the power system from the generator to the end customers. Machine Learning (ML) and Artificial Intelligence (AI) algorithms are used to process and analyze this data. The goal of these systems is to make better judgments on how to run them in the real world.

IoT refers to the installation of sensors in equipment that provide data streams via internet access. "Collect, aggregate, analyze, insight, and act" is a five-step process that all IoT-related services must follow. "Analyze, insight, and act" processes are largely carried out by humans in the absence of artificial intelligence. The majority of the key value from IoT data was lost or unused due to the vast volume of data information that is beyond human comprehension. This is where artificial intelligence technology comes into play in enabling the IoT's usefulness. The term "AIoT" refers to the fusion of established IoT with artificial intelligence technology.

The cloud has been used to store massive amounts of data from traditional industrial equipment. New obstacles arise when it comes to managing and analyzing massive data to get insights. It is critical how the gadget equipped with artificial intelligence is seamlessly connected to the vast database in order to make the devices algorithm-driven. As a result, for an AIoT system, the networking foundation is critical. Furthermore, the equipment contains a great number of sensors, but the quality of the sensors and the absence of original design limit the number of sensors available, resulting in data quality that is subpar.

Overall, industrial equipment has poor data stability, missing data, incompatible formats, a high level of uncertainty, and a high dimensionality.

The cloud has been used to store massive amounts of data from traditional industrial equipment. New obstacles arise when it comes to managing and analyzing massive data to get insights. It is critical how the gadget equipped with artificial intelligence is seamlessly connected to the vast database in order to make the devices algorithm-driven. As a result, for an AIoT system, the networking foundation is critical. Furthermore, the equipment contains a great number of sensors, but the quality of the sensors and the absence of original design limit the number of sensors available, resulting in data quality that is subpar.

Overall, industrial equipment has poor data stability, missing data, incompatible formats, a high level of uncertainty, and a high dimensionality.

It is notably helpful for distributed energy resources systems once a significant number of linked machines and sensors are connected via a closed or worldwide network, as opposed to the previous disconnected generation's operation style. In addition to IoT connection, artificial intelligence enhances the advancement of DER. They can detect faults automatically, optimize system operations autonomously, and forecast future outcomes. A DER system may be characterized as a "smart energy system" using AIoT.

This article presents the definitions of internet of thongs, artificial intelligence and AIoT and their applications on the performances of renewable energies.

After this introduction, Sects. 2 and 3 describe the role of internet of things and artificial intelligence in renewables energies; Sect. 4 reports AIoT Systems and Sect. 5 presents the conclusions.

2 Role of Internet of Things in Renewable Energy

2.1 Internet of Things

The Internet of Things (IoT) is a relatively new concept that intends to connect physical items or "things" over the Internet [1]. Physical devices include household goods as well as industrial machines. These gadgets might give vital data and enable customers to acquire a variety of services if they are equipped with the right sensors and connection networks. For example, decreasing energy expenses can be achieved by intelligently managing building energy use [2]. Manufacturing, transportation, and the construction industry are just a few examples of industries that might benefit from IoT. The IoT has a wide range of applications, including environmental monitoring, energy-efficient building management, services and healthcare systems, and drone-based services [3–7].

The selection of IoT components such as communication protocols, data storage, sensor devices and processing must be appropriate for the desired application when creating an IoT application, which is the initial stage in building IoT systems. Communication technology and environmental sensors, for example, are required by an IoT platform used to control a building's heating, air conditioning and cooling [8]. The components of an IoT platform are depicted in Fig. 1 [9]. Sensors, actuators, IoT gateways, and any other device involved in the data collection, transmission, and processing cycle are examples of IoT devices, which make up the second component of IoT platforms. An IoT gateway device, for example, allows data to be routed into an IoT system in order to create bidirectional connections between the device, the gateway, and the cloud.

The third element of the IoT platform is communication protocols, and they enable diverse devices to connect with controllers or decision-making centers and share data. Based on your application's requirements, IoT platforms provide you the flexibility to pick the type of communication technology you want (each with its own set of capabilities). Wi-Fi, Bluetooth, and ZigBee [10], as well as cellular networks like LTE-4G and 5G networks [11], are examples of these technologies. The IoT platform's data storage component is the fourth component that allows for the management of sensor data.

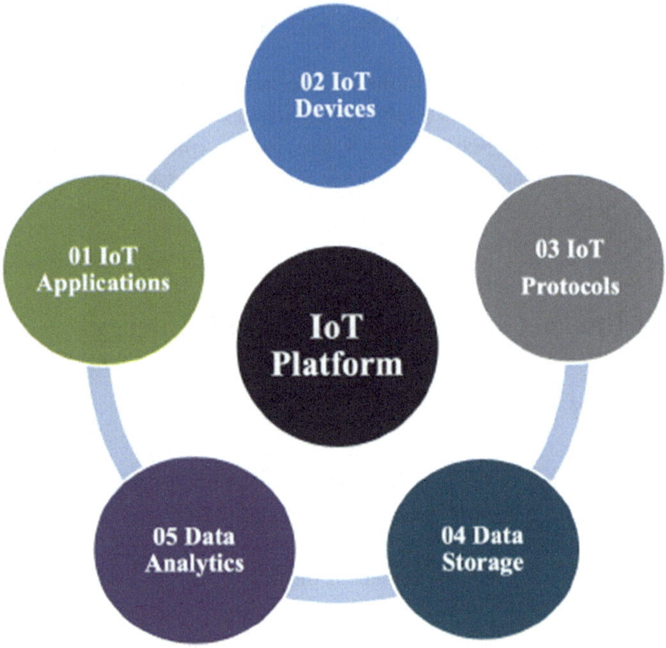

Fig. 1 The components of an IoT platform are depicted in this diagram

The amount of data acquired by the gadgets is enormous in principle. This necessitates developing a data storage method that can be implemented on cloud servers or at the network's edge. The fifth component of IoT platforms is data, which is captured and used for analytical purposes. Data analytics can be performed either offline after the data has been stored or in real time. The purpose of data analysis is to make decisions about the application's functionality. Data analytics can be done in real-time or off-line, depending on the situation. Off-line analytics begins with the collection of stored data, which is subsequently displayed using visualization tools on-premises. Real-time analytics, such as stream analytics [12], rely on cloud or edge servers for visualization.

2.2 *IoT in the Energy Sector*

Fossil fuels now account for roughly 80% of global final energy. Excessive fossil fuel extraction and combustion has detrimental environmental, economic implications and health, such as air pollution and climate change, to name a few. Energy efficiency (using less energy to give the same service) and the utilization of renewable energy sources are two of the most important options for reducing the negative effects of fossil fuel consumption [13, 14].

This section discusses the role of IoT in the energy industry, from fuel extraction through energy-generating asset operation and maintenance (O&M), as well as technology development and end-use. The Internet of Things (IoT) has the potential to cut down on energy waste and carbon emissions. An Internet of Things-based energy management system might track real-time energy use and raise energy performance awareness at every level of the supply chain [15, 16]. The first part of this section discusses IoT implementation in the energy producing phases. Then we'll look at smart cities, which are a catchall word for a variety of IoT-based subsystems including smart buildings, smart industries, smart grids and intelligent transportation.

IoT and Energy Generation

In the 1990s, the power industry grew more interested in automating industrial processes and implementing supervisory control and data gathering systems [17]. Early phases of the Internet of Things began to help the power industry by reducing the risk of output loss or blackout by monitoring and regulating equipment and operations. Old power plants have major problems in terms of reliability, efficiency, environmental effect, and maintenance. Energy losses and unreliability can be caused by the age of equipment in the power sector, as well as inadequate maintenance issues.

Assets might be over 40 years old, extremely valuable, and difficult to replace. Some of these challenges in power plant management can be alleviated with the Internet of Things [17]. IoT sensors can identify any malfunction or unexpected decline in energy efficiency in internet-connected equipment, alerting the need for repair. This increases the system's resiliency and efficiency while cutting maintenance expenses [18]. A new IoT-based power plant might save 230 million dollars over its lifetime, whereas an older facility of the same size could save 50 million dollars, according to [19].

Many nations are supporting RESs as a way to reduce fossil fuel usage and rely on local energy sources. Weather-dependent or variable renewable energy (VRE) sources such as wind and solar energy provide the energy system with additional hurdles described as "the intermittency challenge." Matching energy generation with demand for energy system with a large percentage of VRE is difficult owing to supply and demand fluctuation, which results in mismatch across time scales. IoT systems provide for greater flexibility in balancing generation and demand, reducing the obstacles of installing VRE and resulting in larger clean energy integration shares and lower GHG emissions [20]. Furthermore, by utilizing IoT, more efficient energy consumption may be accomplished through the application of machine-learning algorithms that assist when it comes to selecting the right mix of supply and demand technologies [17]. Artificial intelligence algorithms, for example, can combine numerous small-scale solar PV panels to balance the power output of a thermal power plant [21].

3 Role of Artificial Intelligence in Renewable Energy

3.1 Artificial Intelligence

Artificial Intelligence (AI) is a phrase that, in its widest definition, refers to a machine's or artifact's capacity to perform activities that are similar to those that define human intellect. The term artificial intelligence (AI) refers to computer applications and systems that can accomplish tasks that are more complicated than simple programming, but are still not capable of thinking. Artificial neural networks and expert systems, respectively, are two fields of AI. Expert systems (ES) are logic programs that enable computers to "make judgments" by understanding input and choosing among options. Rule-based inference is a technique used by expert systems, in which input is processed using pre-established rule systems. Regardless of their complexity, systems are still unable to approach the level of intricacy seen in truly intelligent cognition.

The notion of neural network analysis was found approximately 50 years ago, but applications software to handle actual issues has just been created in the recent 20 years. Neural networks' history and theory have been documented in a great amount of published literature, and this paper will only provide a basic summary of how neural networks work. Some jobs are well-suited to ANNs, whereas others aren't. They're especially well-suited to professions that need data sets with gaps, unclear or a lack of information, and a complicated situation and ill-defined circumstances in which employees frequently make judgments based on intuition. They can learn from others' mistakes and tackle non-linear problems.

They are also fault-tolerant and robust. Activities demanding great accuracy and precision, such as logic and mathematics, are tasks that ANNs cannot successfully do.

ANNs have been successfully applied in a wide range of fields, including, neurology, engineering, economics, medicine, mathematics and many more. Pattern, sound, and speech recognition, electromyography and other medical signals analysis, military target identification, and bomb detection in passenger luggage are only a few of the most important. Weather and market trend prediction, mineral exploration site prediction, electrical and thermal load prediction, adaptive and robotic control, and a range of other applications have all used them. In process control, neural networks are used because they can develop predictive models of the process based on multidimensional data collected on a regular basis from sensors.

ANNs have been effectively used in a wide range of domains, including mathematics, engineering, medicine, economics, neurology, and many more. Pattern, sound, and voice recognition, the study of electromyographs and other medical signs, the identification of military targets, and the detection of bombs in passenger baggage are only a few of the most essential. They've also been utilized in weather and market trend forecasting, mineral exploration site prediction, electrical and thermal load prediction, adaptive and robotic control, and a variety of other applications. Neural networks are employed in process control because they can create predictive models of the process using multidimensional data acquired from sensors on a regular basis.

A neural network, according to Haykin [22] is a distributed massively parallel processor with a natural proclivity for storing and making accessible experience information. In two ways, it mimics the human brain: the network acquires information through a learning process, and the knowledge is stored using synaptic weights, which are the strengths of inter-neuron connections. In engineering analysis and forecasts, ANN models might be a viable option. The learning process of a human brain is modeled using artificial neural networks. They work in the same way as a "black box" model, requiring no technical system knowledge. Rather, they analyses previously recorded data to discover the link between input parameters and controlled and uncontrolled variables. Multiple regression analysis can be likened to ANN, except that ANN does not need any assumptions about the system to be simulated. Neural networks have been used to solve a broad range of issues, including non-linear problems that are difficult to solve using traditional approaches. Another benefit of employing ANNs is that they can manage huge and complicated systems with numerous interconnected factors. They appear to simply dismiss irrelevant data and focus on the more critical inputs. Artificial neural networks can discover essential information patterns within a multidimensional information domain without the need of complicated rules and mathematical processes. Furthermore, neural networks are resilient, fault-tolerant, and noise-resistant [23].

To accomplish the least amount of inaccuracy in the quickest amount of time, several techniques are often utilized. There are many distinct types of neural networking systems and many different methods to apply them to a problem. The type of problem to be solved influences the applicability of an acceptable paradigm and method for implementation. Back-propagation (BP) and its derivatives [23] are the most used learning algorithms. The BP method is a gradient descent algorithm and is one of the most powerful learning algorithms in neural networks. It seeks to increase the neural network's performance by lowering the overall error and altering the weights along the gradient. The BP algorithm is described in further depth in [24].

When dealing with an issue, the first step is to gather the necessary information and organize it in a spreadsheet format with multiple columns representing the input and output parameters. A training data file, a test data file, and a validation data file are all necessary. The former and latter should both contain representative samples of all the instances the network is expected to handle, whereas the test file may only contain roughly 10% of the cases in the training file. The network is evaluated against the test file during training, and training should be halted if the mean average error stays unchanged after a certain number of epochs. To solve a problem using ANNs, the right architecture, learning rate, momentum, number of neurons in each hidden layer, and activation function must be chosen. This is a time-consuming and hard operation, but as knowledge is gained, some factors may be anticipated simply, reducing the time necessary.

3.2 Renewable Energy Systems and ANN Applications

Many academics have employed Artificial Neural Networks (ANNs) for modeling and prediction in the field of renewable energy systems. This document organizes a number of such applications by theme. Due to space constraints, additional information about the author's most recent work in the field is provided. Höller et al. [3] provides a more in-depth analysis.

a. **Wind Speed Prediction**

When data isn't readily available Data from three stations from 1986 to 1996 were utilized to train the network, while data from 1997 were used to validate it [25]. The learning and prediction processes were both completed with satisfactory accuracy. Two multilayered artificial neural network topologies of the same sort were tested, one with five neurons in the input layer (month, wind speed at 2 m and 7 m for two stations), and the other with eleven. The meteorological stations' x and y coordinates are used as supplementary input data for the 11-input network. When the unknown data set (year 1997) was provided to the two networks, the greatest percentage difference achieved was 1.8 and 5%, respectively, which is quite excellent.

The two networks may be utilized for distinct tasks: the five-input network can be used to populate a database with missing data, while the eleven-input network can be used to estimate mean wind speed in other adjacent areas. In the first case, the station might be inside the three-station region (interpolation) or outside the area defined by the three stations (extrapolation).

b. **Modelling of a Solar Steam Generator**

The use of Artificial Neural Networks (ANNs) has been used to represent several features of a solar steam generator. A parabolic trough collector, a flash vessel, a high-pressure circulating pump, and the related piping are all used in the system. Here is a description of some of the work done on this system.

The intercept factor is the ratio of absorbed energy by the receiver to energy incident on the concentrator aperture. Collector optical efficiency may be calculated using the intercept factor. This is a critical factor in determining how successful solar concentrating collectors are overall. In comparison to the significantly more sophisticated calculation of the Energy DEPosition (EDEP) computer code [26], ANNs have been able to determine the intercept factor with a difference of less than 0.4%.

On the collector's receiver, the radiation profile is shaped like a "bell." Local concentration ratios at 10° intervals around the receiver's periphery are used to represent this. It's crucial to be able to measure this profile because it allows you to determine the collector's optical efficiency. This test must be performed at various incidence angles, as well as at a normal incidence angle (=0°). Because of the collector's size, this is usually a very difficult task. ANNs have been used to learn the radiation profile from readings at angles where experiments are possible and make predictions for other angles, including the normal incidence angle [27]. The highest discrepancy between ANN forecasts and actual experimental data is 3.2%, which is regarded acceptable.

The system's startup has also been modeled using ANNs [28]. It is critical for the designer of such systems to be able to make such predictions because the energy expended during the morning start-up has a significant impact on the system's performance. Because the system operates under transient conditions, this problem is difficult to solve using analytical methods. The temperature profile at various points of the system could be predicted by ANNs to within 3.9%, which is considered good enough for design purposes. The energy spent during the heat-up period can easily be calculated using the temperature profiles.

The mean monthly average steam generation of the system is an essential metric to consider while designing such systems. A network was trained using performance values for a variety of collection sizes ranging from 3.5 to 2160 m^2 and was able to forecast inside and beyond the training range [21]. In comparison to simulated data, the ANN was able to forecast the mean monthly average steam output with a maximum deviation of less than 5.1%, which is regarded acceptable.

4 AIoT System

AIoT will extend existing IoT standards to create self-contained future communication architectures that will facilitate the intelligent interchange of data between millions of devices. The building design four layers make up the AIoT system: Data analytics, IoT hardware, communication layer, cloud infrastructure (Fig. 2).

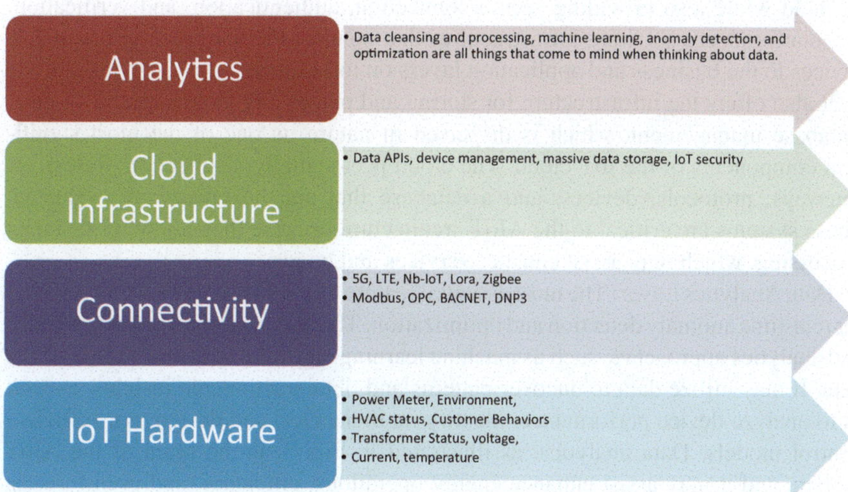

Fig. 2 AIoT system architecture

IoT Gear: Energy assets have their own data source, which requires hardware to connect to. Low-cost IoT sensors may be used to monitor the state of assets like as electricity, flow rate, and pressure. Sensors and actuators are examples of transducers in IoT devices. Traditional power equipment, such as smart light bulbs, linked valves and pumps, smart meters, connected power plants, smart building components, and so on, are becoming connected as a result of low-cost and low-power IoT technology. The platform of AIoT architecture, the cornerstone of linked and intelligent solutions, is data, which is captured by IoT devices. For the top levels, it allows for IoT-related smart applications.

The Connectivity Layer is in charge of acquiring and transmitting data over a network. Sensors and devices are the major components that make up the connection layer. Sensors gather data and convey it to the next layer, where it is processed. As technology advances, semiconductor technology is being leveraged to create miniature smart sensors that may be employed in a variety of applications.

IoT systems' backbone is the communication layer. It connects the application layer with the IoT hardware layer. The AIoT system has a lot of data and information that has to be communicated. As a result, a low-cost, low-bandwidth network must be established between these nodes. All industrial equipment communication protocols, such as Modbus and DNP3, must be interpreted by the communication layer, which can transport data over wired or wireless protocols like Lora, NB-IoT, and 5G. As a result, this is a large layer that encompasses a variety of devices, technologies, solutions (both software and hardware), and functions. IoT gateways are often used for connection aggregation, protocol translation, IoT data encryption and decryption, device administration, and sophisticated edge computing. Furthermore, in order to link objects to the outside world, networks are critical components in AIoT.

Cloud infrastructure must be able to handle a high number of IoT endpoints in the field while also providing secure connection, authentication, and verification. IoT platforms are software infrastructures that connect the hardware layers of IoT devices to the business and application layers on top of them.

It also offers the infrastructure for storing and processing large amounts of data. Database management, which is dispersed in nature, is one of the most significant components of the IoT cloud. The cloud is essentially a collection of devices, gateways, protocols, devices, and a database that can be effectively evaluated. These systems are critical to the AIoT architecture because they allow for fast data processing, which may assist enhance services and goods.

Data Analytics Layer: The most essential feature of IoT technology is that it allows for real-time anomaly detection and optimization. This layer uses various data science and analytics approaches, such as machine learning algorithms, to make sense of the data. It may utilize data to uncover patterns and obtain actionable insights, as well as to analyze device performance, identify inefficiencies, and design more efficient control models. Data analytics, as illustrated in Fig. 3, is the brain of the AIoT system, and it may assist enhance energy operations, efficiency, and even forecast future occurrences such as equipment failure.

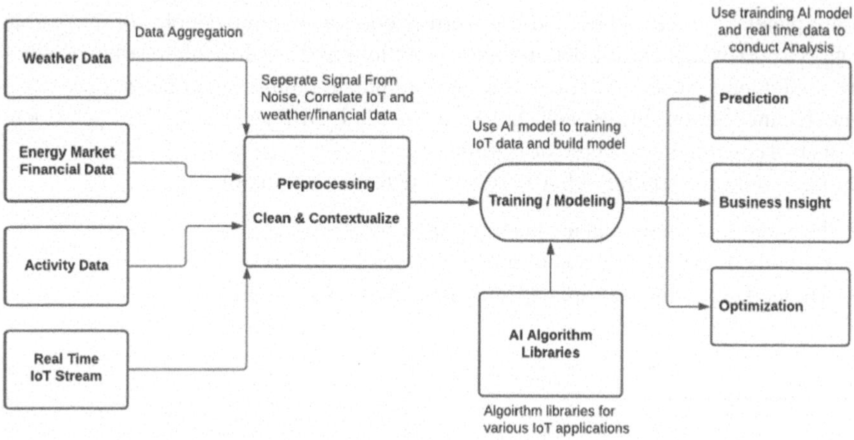

Fig. 3 AI data process flow for IOT application

4.1 AIoT's Advantages in Distributed Energy Resources

The Internet of Things brings a slew of benefits to energy systems, including preemptive actions and real-time intelligent automation, boosting the renewable energy industry's production and profitability.

a. **Increasing the Efficiency of Operations**

To enhance renewable energy generating efficiency and capacity, accurate and real-time data analytics is essential. AIoT will assist renewable energy firms in managing the vast capacity of assets situated in remote and geographically dispersed places, which can be difficult to manage. Using powerful algorithms, AIoT can examine asset data in real time. Machine learning combined with AI can also forecast operational circumstances. It is possible to employ technology to ensure that energy transformation occurs in a uniform manner. AIoT can detect changes in demand and supply and automatically deliver information to operators so they can respond to these changes and boost efficiency.

As a result, AIoT can reveal which operations are redundant and time-consuming, allowing renewable energy assets' productivity to be increased.

b. **Better Production Control**

In Distributed Energy Systems, decisions concerning energy generation, load switching, and network design modifications are made on a regular basis. AIoT data analytics can give useful information for predicting electricity generation and asset dependability. For example, AIoT provides real-time access to the operating status of a solar farm, which can be translated into actionable insights, foresee a wide variety of dangers, and automate for quick reaction.

c. **Costly Downtime is Reduced.**

Solar panels, wind turbines, and transformer equipment are examples of renewable energy sources that are located in remote locations and are difficult to reach. Predictive maintenance using AIoT can forecast equipment failure and organize preventive maintenance actions in advance. As a result of AIoT, DERs can minimize operational costs by reducing unexpected downtime.

According to Deloitte, AIoT produces the following outcomes:

- They cut their maintenance planning time by 20–50%.
- Equipment availability and uptime increased by 10–20%.
- Maintenance expenses are expected to decrease by 5–10%.

5 Conclusion

The definitions of the internet of things, artificial intelligence, and the internet of things are presented in this article, as well as their implications for renewable energy and energy efficiency. Microgrids face significant problems from dispersed energy resources such as solar panels, storage, and power transformers. Artificial intelligence and the Internet of Things are combining to generate new potential in the realm of distributed energy resources.

References

1. Haseeb, K., Almogren, A., Islam, N., Ud Din, I., Jan, Z.: An energy-efficient and secure routing protocol for intrusion avoidance in IoT-based WSN. Energies **12**, 4174 (2019)
2. Zouinkhi, A., Ayadi, H., Val, T., Boussaid, B., Abdelkrim, M.N.: Auto-management of energy in IoT networks. Int. J. Commun. Syst. **33**, e4168 (2019)
3. Höller, J., Tsiatsis, V., Mulligan, C., Avesand, S., Karnouskos, S., Boyle, D.: From Machine-to-Machine to the Internet of Things: Introduction to a New Age of Intelligence. Elsevier, Amsterdam (2014)
4. Atzori, L., Iera, A., Morabito, G.: The internet of things: a survey. Comput. Netw. **54**, 2787–2805 (2010)
5. Hui, T.K., Sherratt, R.S., Sánchez, D.D.: Major requirements for building smart homes in smart cities based on internet of things technologies. Future Gener. Comput. Syst. **76**, 358–369 (2017)
6. Evans, D.: The internet of things: how the next evolution of the internet is changing everything. CISCO White Pap. **1**, 1–11 (2011)
7. Motlagh, N.H., Bagaa, M., Taleb, T.: Energy and delay aware task assignment mechanism for UAV-based IoT platform. IEEE Internet Things J. **6**, 6523–6536 (2019)
8. Ramamurthy, A., Jain, P.: The Internet of Things in the Power Sector: Opportunities in Asia and the Pacific.; Asian Development Bank, Mandaluyong, Philippines (2017)
9. Jia, M., Komeily, A., Wang, Y., Srinivasan, R.S.: adopting internet of things for the development of smart buildings: a review of enabling technologies and applications. Autom. Constr. **101**, 111–126 (2019)
10. Karunarathne, G.R., Kulawansa, K.T., Firdhous, M.M.: Wireless communication technologies in internet of things: a critical evaluation. In: Proceedings of the 2018 International Conference on Intelligent and Innovative Computing Applications (ICONIC), Plaine Magnien, Mauritius, 6–7 December 2018, pp. 1–5 (2018)

11. Li, S., Da Xu, L., Zhao, S.: 5G internet of things: a survey. J. Ind. Inf. Integr. **10**, 1–9 (2018)
12. Watson Internet of Things. Securely Connect with Watson IoT Platform (2019). https://www.ibm.com/internet-of-things/solutions/iot-platform/watson-iot-platform. Accessed 15 Oct 2019
13. Connolly, D., Lund, H., Mathiesen, B.: Smart Energy Europe: the technical and economic impact of one potential 100% renewable energy scenario for the European Union. Renew. Sustain. Energy Rev. **60**, 1634–1653 (2016)
14. Grubler, A., Wilson, C., Bento, N., Boza-Kiss, B., Krey, V., McCollum, D.L., Rao, N.D., Riahi, K., Rogelj, J., De Stercke, S., et al.: A low energy demand scenario for meeting the 1.5 C target and sustainable development goals without negative emission technologies. Nat. Energy **3**, 515–527 (2018)
15. Tan, Y.S., Ng, Y.T., Low, J.S.C.: Internet-of-things enabled real-time monitoring of energy efficiency on manufacturing shop floors. Procedia CIRP **61**, 376–381 (2017)
16. Bhardwaj, A.: Leveraging the Internet of Things and Analytics for Smart Energy Management. TATA Consultancy Services, Mumbai (2015)
17. Ramamurthy, A., Jain, P.: The Internet of Things in the Power Sector: Opportunities in Asia and the Pacific. Asian Development Bank, Mandaluyong, Philippines (2017)
18. Sigfox, Inc. Utilities & Energy (2019). https://www.sigfox.com/en/utilities-energy/. Accessed 27 Sept 2019
19. Immelt, J.R.: The Future of Electricity Is Digital; Technical Report. General Electric, Boston, MA (2015)
20. Al-Ali, A.: Internet of things role in the renewable energy resources. Energy Procedia **100**, 34–38 (2016)
21. Karnouskos, S.: The cooperative internet of things enabled smart grid. In: Proceedings of the 14th IEEE International Symposium on Consumer Electronics (ISCE2010), Braunschweig, Germany, 7–10 June 2010, pp. 7–10 (2010)
22. Haykin, S.: Neural Networks: A Comprehensive Foundation. Macmillan, New York (1994)
23. Rumelhart, D.E., Hinton, G.E., Williams, R.J.: In: Parallel Distributed Processing: Explorations in the Microstructure of Cognition. MIT Press, Cambridge, MA (1986)
24. Kalogirou, S.A.: Renew. Sustain. Energy Rev. **5**, 373 (2001)
25. Kalogirou S.A., Neocleous C., Paschiardis S., Schizas C.: In: Proceedings of the European Symposium on Intelligent Techniques on CD-ROM, Crete, Greece (1999)
26. Kalogirou, S.A., Neocleous, C.C., Schizas, C.N.: In: Proceedings of the International Conference on EANN'96, London, UK, pp. 5–8 (1996)
27. Kalogirou S.A.: In: Proceedings of Eurosun'96, Freinberg, Germany, pp. 470–475 (1996)
28. Kalogirou S.A., Neocleous C.C., Schizas C.N.: In: Proceedings of the International Conference EANN'96, London, UK, pp. 1–4 (1996)

Design of a Switching Module for Electricity Supply from Solar to Grid by Sensing Light Intensity

Sajal Naz, Sheza Yasin, Abdul Rauf Bhatti, Sarosh Ahmad, M. Usman Siddiqi, and M. Faizan Saleem

Abstract Electricity is the main thrust in a country's economy and the emerging world desires consistent low-cost electricity for its rapid growth. This paper examined how an automatic switching of photovoltaic (PV) array for electricity supply from solar to grid using sensing light intensity. If the intensity of sunlight is optimum for the required load, then PV power will be used and extra power will be stored by the energy storage unit (ESU). If sunlight intensity is not sufficient it will move directly on the grid instead of the usage of energy from ESU. The result shows that the proposed system directly depends on the intensity of sunlight. Our system will ensure longtime availability of energy than the previous system. Moreover, this system can also be utilized in remote places where there is no grid accessible. The proposed system was tested and implemented that had shown excellent results.

Keywords Solar to grid · Switching module · Sensing light intensity · Photovoltaic · Energy storage unit (ESU)

1 Introduction

The country's economic growth heavily builds upon the long-time attainability of energy through the low set and attainable resources. Renewable energy is referred to as clean energy which is environment friendly and less costly that can be obtained from natural resources or processes that are continuously replenished. The analysis of

S. Naz · S. Yasin · A. Rauf Bhatti · S. Ahmad (✉) · M. Usman Siddiqi · M. Faizan Saleem
Department of Electrical Engineering and Technology, Government College University Faisalabad (GCUF), Faisalabad 38000, Pakistan
e-mail: saroshahmad@ieee.org

A. Rauf Bhatti
e-mail: bhatti_abdulrauf@gcuf.edu.pk

S. Ahmad
Department of Signal Theory and Communications, Universidad Carlos III de Madrid, 28911 Leganes, Madrid, Spain

S. El Himer et al. (eds.), *Artificial Intelligence of Things for Smart Green Energy Management*, Studies in Systems, Decision and Control 446,
https://doi.org/10.1007/978-3-031-04851-7_2

electric vehicles committing through Energy Storage Unit and PV coupled to microgrid. This analysis for electric vehicles committing through Energy Storage Unit (ESU) and PV coupled to microgrid. This system is built on heuristic regulation-based policies and different operating modes to control the flow of energy in microgrid has shown that a scheme regulation-based energy management for charging of electric vehicles from Photo-voltaic (PV) grid-connected system [1–3]. In several cases, photovoltaic grid connected systems do not normally have battery accumulator; additional energy from the Photovoltaic array sent to the utility grid later delivering the main load. On the contrary, energy drained from the grid when the energy produced from PV is inadequate for the main load. Consequently, there is a difference between the cost of buying electricity from the grid and the selling price. The photovoltaic system's grid is a significant ingredient in identifying the suitable size of Photovoltaic grid-connected systems [2–4]. Photovoltaic-based reserves system in low grid dependability areas. This system excerpts greatest possible power from photovoltaic dynamo by using a simple Maximum Power Point Tracking technique which is implemented through a cheap 8 bits microprocessor [6]. The constant price charging without any interruption is obtained by managing flow of energy between grid ESU and PV. According to the rules which are defined by a scheme regulation-based energy management [7]. A report that describes six year's results, to the condensation of the costs of operation and maintenance (O&M) of $6.3 million project at power plants where the technology of concentrated solar power (CSP) is employed in [8]. Automatic streetlights do not require a standard switching on and off operation. The system observes apart whether light is needed or not. when dullness falls at a constant level, the streetlight is turned on axiomatically. And when there is another cause of light, the streetlight switch is off [9]. The energy flow diagram between photovoltaic and energy storage unit. Energy Flow between Grid ESU & PV is shown in Fig. 1.

The grid means that the solar system is connected to the system of the local benefit company. Grid systems are connected to electricity benefits, which set up the home to use electricity produced by the solar system also as power from limited benefits. This system can turn the electricity back into the grid at low acceptance and high-level production. Actually connected to the grid is helpful as we do not have to purchase an overprice reserves system to keep overflow energy [11]. The program that gives the primary idea of controlling the intensity of streetlights through a microprocessor also as saving energy, as a result of which the programming microprocessor was occupied in providing the different light intensity in different situations [12]. In [13], the automatic design presented switching system that is an automatic conventional manual design switching system. Automated systems are always better Man-made switching can also make a person forget But do not forget about the right time of a machine. In addition to additional salary costs, switching or the operator's pay is also in the manual system. Conventional Photovoltaic power is the maximum ratio of small to medium off-thunder power [14, 15].

The proposed system was tested and implemented that had shown excellent results. The result shows that the proposed system directly depends on the intensity of sunlight. Our system will ensure long-time availability of energy than the previous

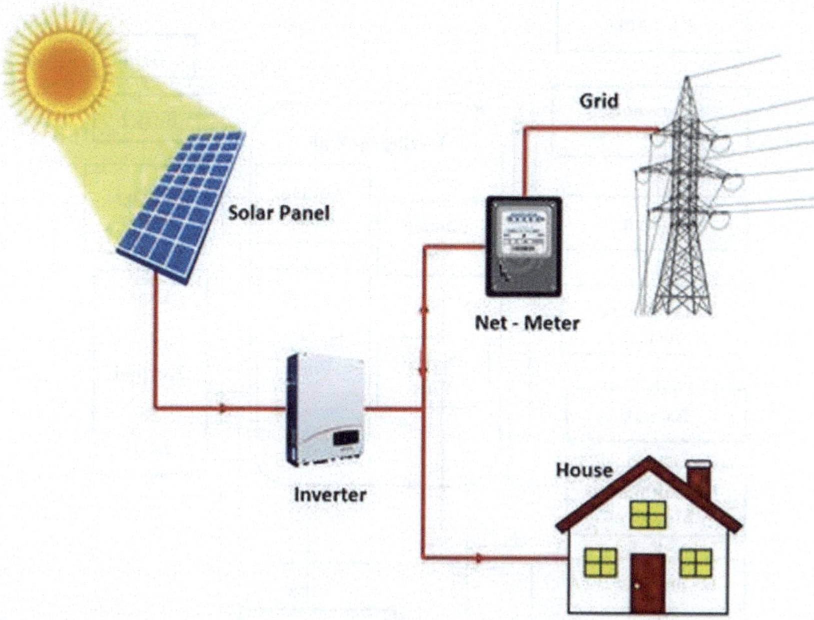

Fig. 1 Energy flow between Grid ESU & PV. [10]

system. Moreover, this system can also be utilized to remote places where there is no grid accessible.

2 Proposed Methodology

This research proposed the automatic switching of PV array by sensing the intensity of sunlight directly. Nowadays the commercially available systems depend upon the level of energy stored in energy storage units (ESU) i.e., batteries. If stored energy is insufficient to entertain the load, then the system switches the load at a grid. The proposed system depends directly on the intensity of sunlight. If the intensity of sunlight is optimum for the required load, then PV power will be used and the extra power will be stored by ESU. If sunlight's intensity is not sufficient then it will move directly on the grid instead of the usage of energy from ESU. The stored energy of ESU will be used if the grid is not available and the sunlight is absent. This system will ensure the longtime availability of energy than previous systems. Block Diagram of Automatic Switching of PV is shown in Fig. 2.

We also have provided a block diagram of the entire system which is presented in Fig. 3.

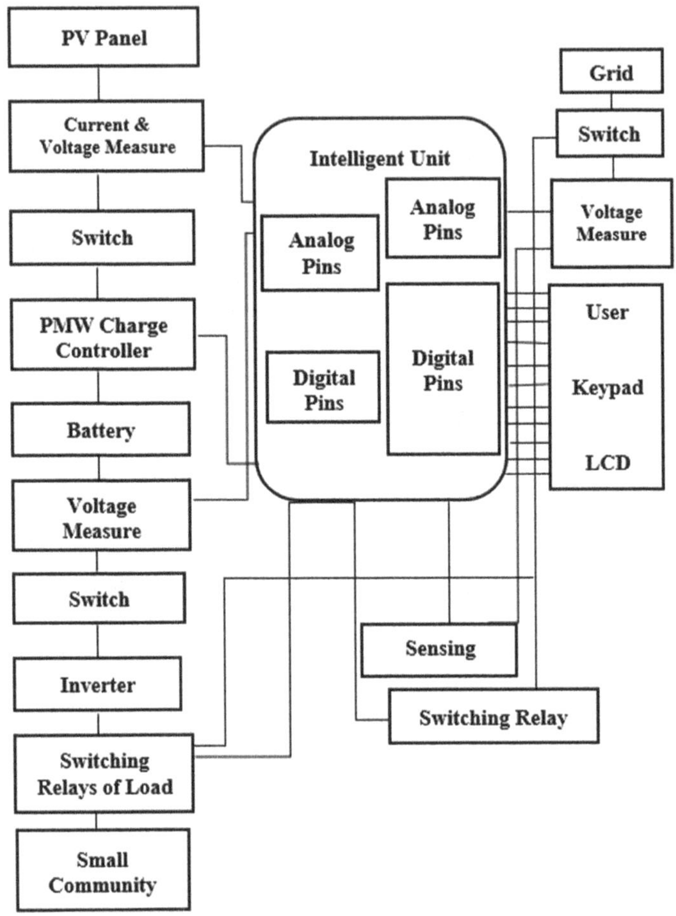

Fig. 2 Block diagram of automatic switching of PV

2.1 Magnetic Contractor

One of the main electrical parts which can control power devices. Contractors is an electrically controlled switch relay used for switching in an electrical power circuit. The contractor is controlled by a circuit with a low level of the switched circuit. It is an electrically switch for the switching of power circuit-like relays for a high current rating. Mostly used with lightning, distribution panels, motors, and Transformers. Circuit Diagram of the Magnetic Contractor can be seen in Fig. 4.

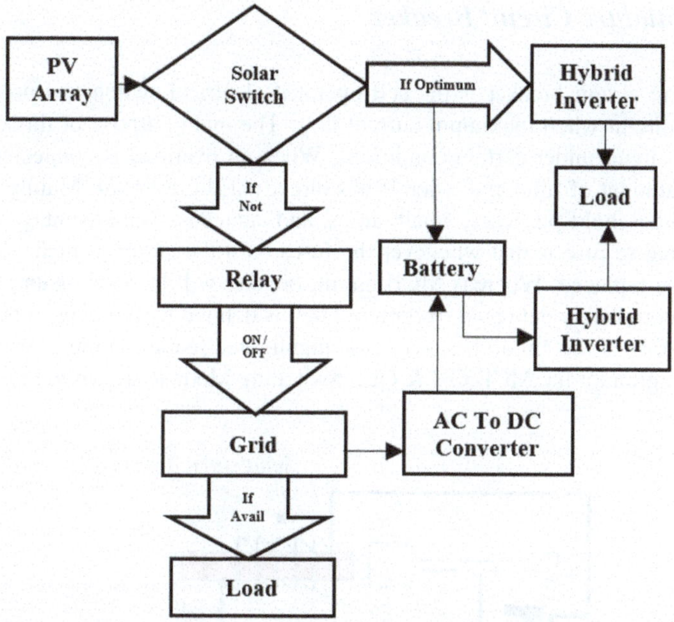

Fig. 3 Block diagram of proposed scheme

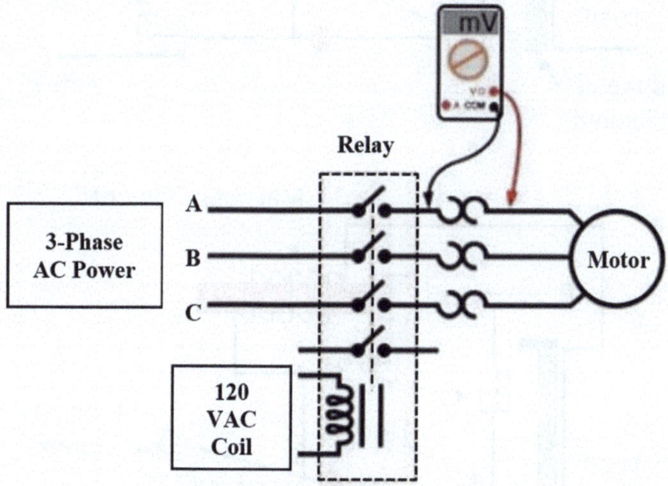

Fig. 4 Circuit diagram of the magnetic contractor

2.2 *Miniature Circuit Breaker*

A miniature circuit breaker is the self-operated electrical switch and protects the electrical circuit when maximum current flow. The main purpose of this device to break the circuit under certain conditions. When an overload is connected during a certain amount of time and when it has directly a short circuit. Mainly used for industry, household purposes, small shops, and canteens. The advantage of MCB as compared to fuse is that whenever the fuses burn the circuit it melts down and needs to be replaced. Whereas MCB circuit breaker will be used again and again until it is completely damaged or ruptured that is defined by breaking capacity. So, the reusability of the circuit breaker is the highlighted feature in these two devices. Circuit Diagram of the MCB ON & OFF Switching Mode is shown in Fig. 5.

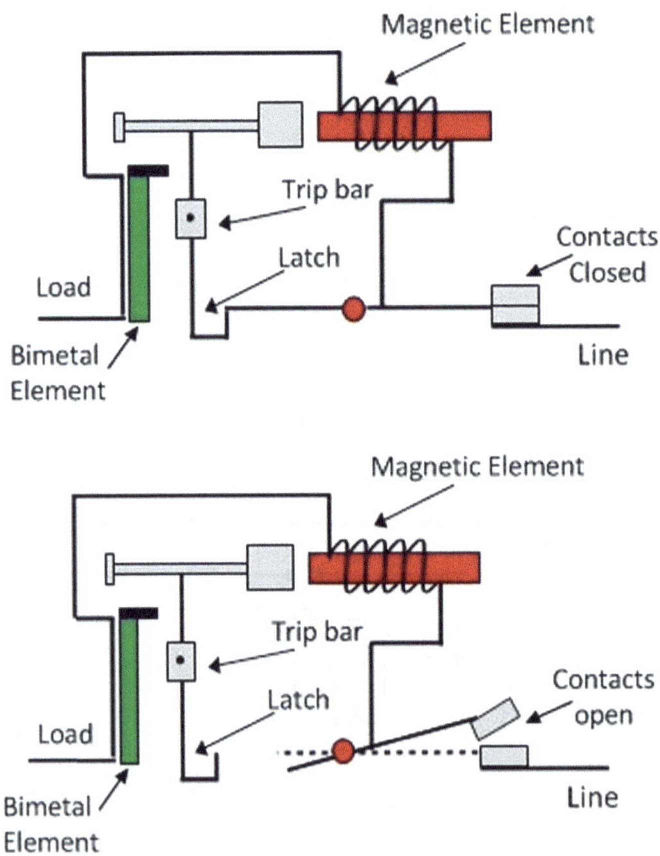

Fig. 5 Circuit diagram of the MCB ON & OFF switching mode

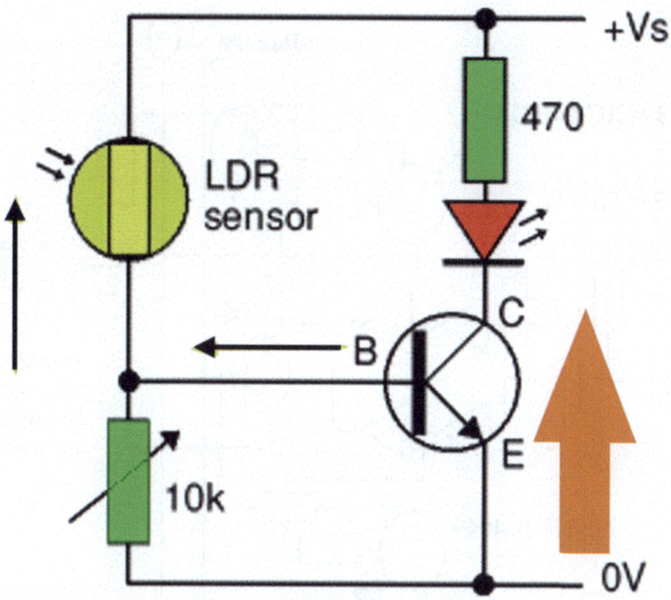

Fig. 6 Circuit diagram of the temperature and light Sensor/Solar switch

2.3 *Temperature and Light Sensor/Solar Switch*

Light Sensor/Solar switch are photo-electric devices. Basically, this is the PIR (passive infrared sensor) which can be automatic on/off the devices through light and also convert light energy into infrared or visible light into an electrical signal. They are used to sense the intensity of sunlight and temperature respectively. It senses the intensity of light optimum or not optimum. Circuit Diagram of the Temperature and Light Sensor/Solar switch as shown in Fig. 6.

2.4 *Solar Hybrid Inverter*

A simple inverter converts DC into AC while a hybrid inverter goes a step further and works with batteries to store excess power as well. They solve issues of renewable energy variability and unreliable grid structure. This inverter run consider of solar, normal power and other is power bank. If only use solar it is possible to run hybrid inverter and solar another, if use normal power it is possible to run hybrid inverter and power and solar close then hybrid inverter run consider or battery power handle. Circuit Diagram of the Solar Hybrid Inverter is presented in Fig. 7.

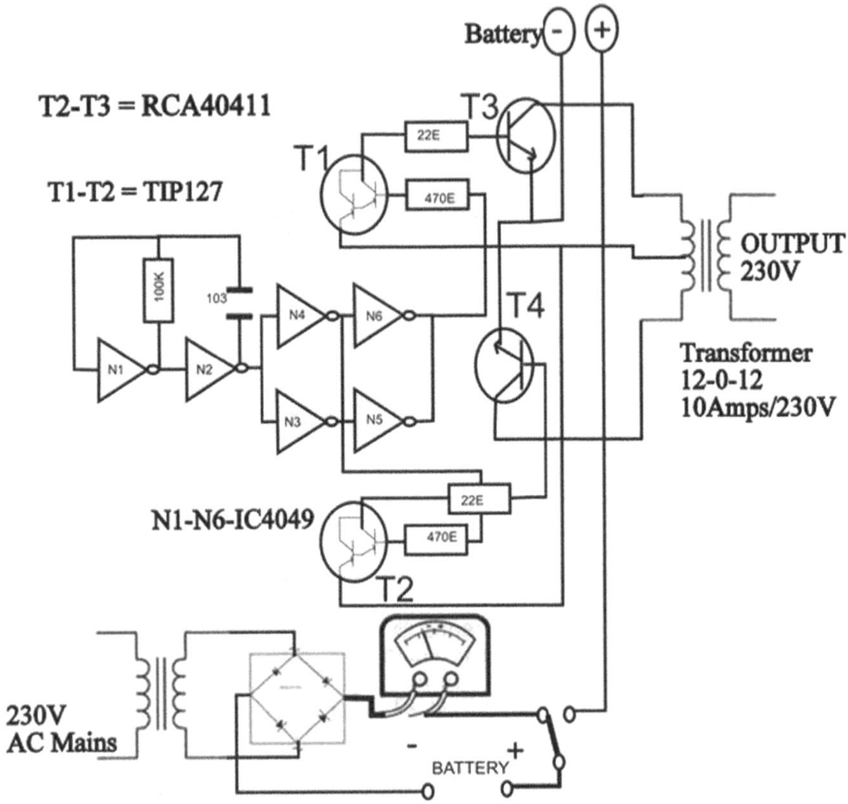

Fig. 7 Circuit diagram of the solar hybrid inverter

2.5 *Photovoltaic Modules*

Solar cell scaled on a structure and inter-connected to form a module. Photovoltaic (PV) modules provide electricity from sunlight. Solar panels have different voltage ratings. Those with 12 or 24 V are best for off-grid systems with battery banks. Other panels come in less common nominal voltages such as 18, 42 and even 60 V. These are used in grid-tied inverters. PV module mounting systems include hardware to permanently affix the array to either a roof, a pole or the ground.

2.5.1 The AC Relay

When the relay coil becomes strong, the current flowing from the coil creates a magnetic field. Like in a DC unit where the polarity is set, or in an AC unit where the polarity changes 120 times per-second, the purpose remains the same: the magnetic coil turns on a ferrous plate, which Part of the structure. A relay is an electromagnet that has a lever to change the electrical charge from the inner spring. To strengthen the relay, the current, alternating current (AC) or direct current (DC) passes through the coil of the electromagnet, magnetizing the core to turn on the lever. The AC relay was cut off due to macular inertia and minor mechanical hysteresis, and the fact that both the North and South Poles coincided. It works on the concept of electromagnetic attraction. When an error is detected in the relay circuit, it reinforces the electromagnetic field which make a temporary magnetic field. This magnetic field transmits the relay armature to open or close the links. Relays are used to serve time delay functions. They are used for delays and delays in getting close to contacts. In relay low voltage signals control the high voltage circuits like low current signals control the high current circuits: (a) Circuit Diagram of the Relay, (b) Types of Relays used in our Proposed System can be seen in Fig. 8.

2.5.2 Solar Switch Timer

Timer circuits are used to create intervals to activate the load. This time the delay is fixed by the user. With the timer working, the time switch controls the out-of-control and off-time accordingly. With the pulse output working, the time switch releases the pulse of the fixed time width at the define/limit time. Turn on the power with the time switch. Before the employers come to work, warmly start the work now/instantaneously/without delay. Figure 9a presents the Circuit Diagram of the Solar Switch Timer, and the Fig. 9b presents the Proposed Solar Switch.

3 Results and Discussion

The purpose of the automatic switching of PV array by sensing Intensity of sunlight, energy store in energy storage unit (ESU) i.e., batteries. If stored energy is insufficient to entertain the load, then system Switches the load at the grid. If the intensity of sunlight is optimum for required load, then PV power will be used and the extra power will be stored by ESU. If sunlight intensity is not sufficient then it will move directly on grid instead of usage of energy from ESU. The stored energy of ESU will be used if grid is not available & sunlight is absent. We are aiming to design a module that will switch supply of electricity from solar to grid by sensing light intensity using sunlight If sunlight is not sufficient to fulfill our needs of energy. Here is working of our project when we will apply our solar supply through Miniature Circuit Breaker (MCB) & our circuit will be turn on.

A miniature Circuit Breaker is electrical safety breaker that break the circuit in two main cases:

- When we have directly a short-circuit.
- And when overload connected during certain amount of time. AC relay switches our circuit through timer.

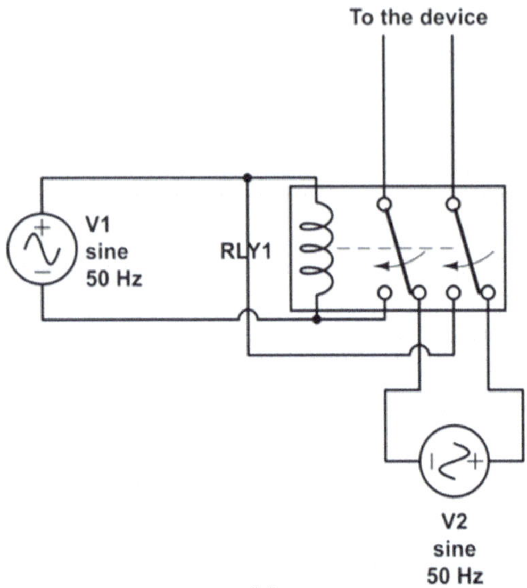

(a)

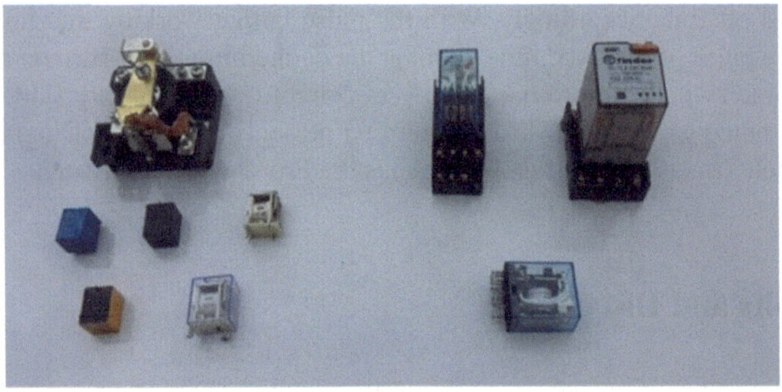

(b)

Fig. 8 **a** Circuit diagram of the relay, **b** Types of relays used in our proposed system

We use timer to remove the residual current from the circuit. Timer circuits are used to create intervals to activate the load. After given time current will be through the magnetic contactor. But magnetic contactor is controlled by solar switch. When the applied control voltage is equal to the rated voltage of the coil, it will strengthen the coil. Whenever the coil is energetic, it creates a magnetic field. Interference with this produces an iron core on the electromagnetic output and then attracts to another coil attached to the plunger. When it attracts the core towards itself the plunger moves down, and it will short all the main contact across each other and after that an electrical machine will be supplied an electrical power this contactor. At the same time, their position will change with the help of auxiliary contact. Normally the close connection will now be open, and the normally open contact will now get closer using

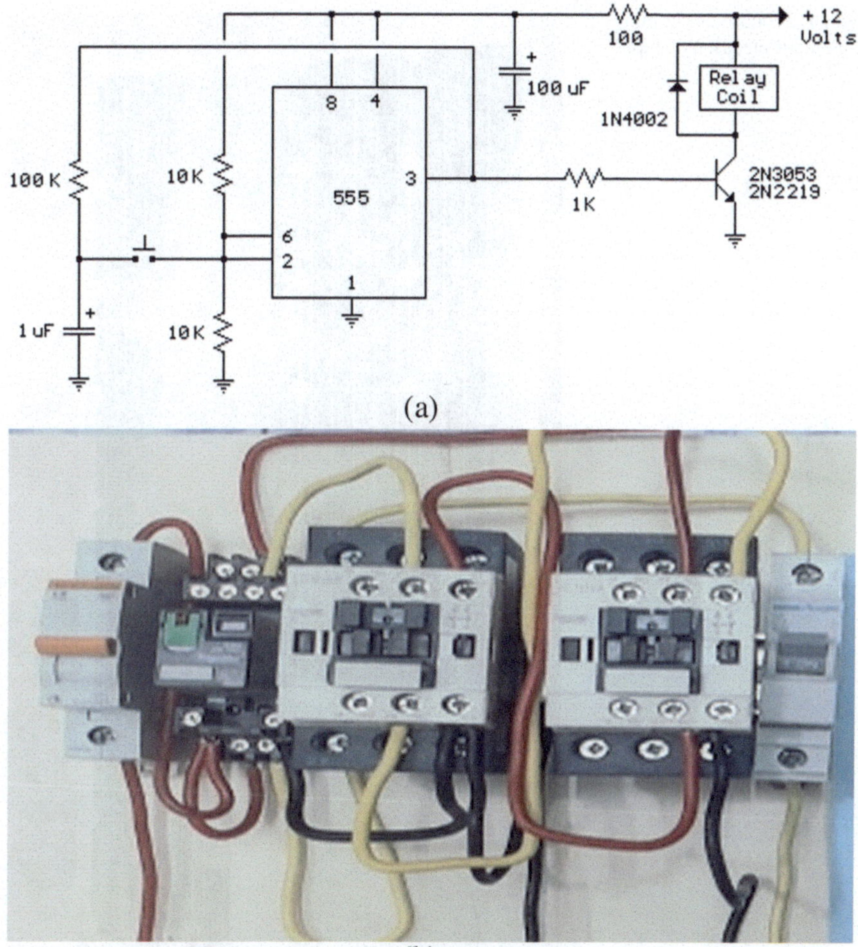

Fig. 9 **a** Circuit diagram of the solar switch timer, **b** Proposed solar switch

this auxiliary contact. The whole circuit of magnetic perform will perform different functions. Proposed Switching Module for Electricity Supply as shown in Fig. 10.

Solar switch senses light intensity if optimum & thorough Switch magnetic flux is produced in coil of contactor & its electricity to load. If not, optimum intensity provides then no magnetic flux produces in Contactor then WAPDA will be used & switch automatically through AC relay and then overload will be entertained by WAPDA. Auto change over relay is used as optional. A relay is an electromagnetic switch that operates through a relatively small electric current that can turn on or off a large electric current. The heart of a relay is an electromagnet (a coil of wire becomes a temporary magnet when electricity passes through it). During switching time our circuit remain off if we wait to entertain load also during switching time. Then our

Fig. 10 Proposed switching module for electricity supply

auto relay which is connected through an MCB will switch our load solar supply during switching time. Hence in this way we made possible long-time availability of electricity. Hardware Setup in is presented in Fig. 11.

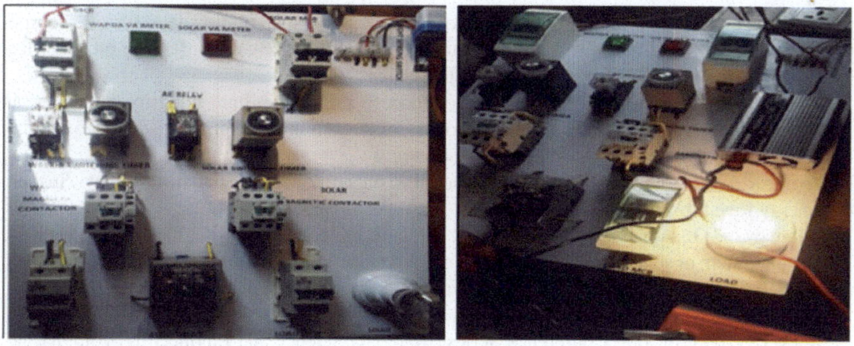

Fig. 11 Hardware setup of the proposed system

4 Conclusion

This paper examined how an automatic switching of photovoltaic (PV) array for electricity supply from solar to grid using sensing light intensity. If the intensity of sunlight is optimum for the required load, then PV power will be used and extra power will be stored by the energy storage unit (ESU). If sunlight intensity is not sufficient it will move directly on the grid instead of the usage of energy from ESU. The result shows that the proposed system directly depends on the intensity of sunlight. Our system will ensure long-time availability of energy than the previous system. Moreover, this system can also be utilized to remote places where there is no grid accessible. The proposed system was tested and implemented that had shown excellent results.

References

1. Baldonado, M., Chang, C.-C.K., Gravano, L., Paepcke, A.: The stanford digital library metadata architecture. Int. J. Digit. Libr. **1**, 108–121 (1997)
2. Bhatti, A., Salam, Z., Ashique, R.H.: Electric vehicle charging using photovoltaic based microgrid for remote Islands. Energy Proc. **103**, 213–218 (2016)
3. Wu, T., Zhang, H., Shang, L.: Optimal sizing of a grid-connected hybrid renewable energy systems considering hydroelectric storage. Energy Sources Part A: Recover. Util. Environ. Eff. 1–17 (2020)
4. Yakuphanoglu, F.,Tang, K.-H., Chao, K.-H., Chao, Y.-W., Chen, J.-P.: Design and implementation of a simulator for photovoltaic modules. Int. J. Photoenergy **368931** (2012)
5. Bhatti, A., Salam, Z.: A rule-based energy management scheme for uninterrupted electric vehicles charging at constant price using photovoltaic-grid system. Renew. Energy **125** (2018)
6. Iqbal, A., Tariq Iqbal, M.: Design and analysis of a stand-alone PV system for a rural house in Pakistan. Int. J. Photoenergy **4967148** (2019)
7. Brhane, Y.: Design & simulation of solar-powered automatic street lighting for Adigrat University. Int. J. Sci. Eng. **11**(8) (2020)

8. Kumar, N.M., Subathra, M.S.P., Edwin Moses, J.: On-Grid solar photovolatic system: components, design considerations, and case study. In: 4th International Conference on Electrical Energy Systems (ICEES), pp. 616–619 (2018)
9. Rout, M.K., Dhar, J., Meher, S.: Design of modern solar street light intensity controller: an energy saving approach. In: 2nd International Conference on Emergrnt Trends in Computing and Communication (2015)
10. Ayop, R., Tan, C.W.: A comprehensive review on photovoltaic emulator. Renew. Sustain. Energy Rev. **80**, 430–452 (2017)
11. Rahmawati, D., Aprillia, B., Priharti, W., Silalahi, D., Kumillayla, K.: Design of automatic switch system of solar panel and power plant for residential load using artificial neural network. In: IOP Conference Series: Materials Science and Engineering, p. 771 (2020)
12. Rinaldi, R., Aprillia, B.S., Ekaputri, C., Reza, M.: Design of open loop single axis solar tracker system. In: IOP Conference Series: Materials Science and Engineering, vol. 982, no. 1, p. 012016. IOP Publishing (2020)
13. Alagumeenaakshi, M., Umamaheswari, S., Mevis, A.A., Seetha, S., Hema, G.: Monitoring and controlling of solar photovoltaic cells using LoRa technology. In: IEEE International Conference on Advancements in Electrical, Electronics, Communication, Computing and Automation (ICAECA), pp. 1–6 (2021)
14. Manfaluthy, M., Pangestu, A., Arif, R., Sanjaya, L.A.: Watt peak meter of solar panel. In: Journal of Physics: Conference Series, vol. 2019, no. 1, p. 012097. IOP Publishing (2021)
15. Makkiabadi, M., Hoseinzadeh, S., Taghavirashidizadeh, A., Soleimaninezhad, M., Kamyabi, M., Hajabdollahi, H., Majidi Nezhad, M., Piras, G.: Performance evaluation of solar power plants: a review and a case study. Processes **9**(12), 2253 (2021)

Nonlinear Backstepping Control of a Grid-Connected Doubly Fed Induction Generator Wind Turbine

Elmostafa Chetouani, Youssef Errami, Abbellatif Obbadi, and Smail Sahnoun

Abstract This chapter suggests a nonlinear Backstepping Control approach to manage a variable wind turbine system. The generator used is a dual-fed induction generator (DFIG). The DFIG's stator is instantly linked to the electrical grid. However, its rotor is connected to the electrical grid through a bidirectional Back-to-Back converter. The proposed methodology, which is based on the stability of the Lyapunov Function, is employed to generate a pulse width modulation (PWM) for the converters. The chapter aims to manage the active and reactive power independently that are injected into the power system. Then again, a maximum power point tracking strategy is used to gain the highest extracted power of a variable wind velocity profile. The introduced control is compared to a vector control strategy. The Simulation results by dint of Matlab/Simulink environment demonstrate that the recommended methodology ensures better system findings in terms of stability, static error, and overshoot.

Keywords Backstepping · DFIG · DFOC · IFOC · GSC · Lyapunov function · MPPT · RSC · WECS · Wind turbine

1 Introduction

In recent years, there has been a significant advancement in the production and utilization of renewable energies. Wind power is one of these alternative sources of energy because it generates electricity from the kinetic energy of the variable

---

E. Chetouani (✉) · Y. Errami · A. Obbadi · S. Sahnoun
Laboratory Electronics, Instrumentation and Energy – Team: Exploitation and Processing of Renewable Energy, Department of Physics, Faculty of Sciences, University of Chouaib Doukkali, El Jadida, Morocco
e-mail: mschetouani@gmail.com

Y. Errami
e-mail: errami.y@ucd.ac.ma

A. Obbadi
e-mail: obbadi.a@ucd.ac.ma

S. El Himer et al. (eds.), *Artificial Intelligence of Things for Smart Green Energy Management*, Studies in Systems, Decision and Control 446,
https://doi.org/10.1007/978-3-031-04851-7_3

wind. This energy is essential to the success of an ecologically sustainable, low-carbon economy. Several studies in electrical, electromechanical, and mechanical engineering have been conducted on this source of energy in an attempt to optimize and increase the quality of the generated power [1]. The principal components of the wind energy system are the wind turbine and the generator. The latter can be classified into three categories: Direct Current (DC), Alternating Current (AC) synchronous, and Alternating Current (AC) Asynchronous generators. The most used types of generators used in the wind application are doubly-fed induction generators (DFIGs), squirrel cage induction generators (SCIGs), and permanent magnet synchronous generators (PMSGs) [2]. Each kind of generator can be run at a fixed or adjustable speed. Because wind output fluctuates, it is best to operate the WTG at variable speed, which lowers physical stress on the turbine blades and drive train while also improving system aerodynamic efficiency and torque transient performance [3]. Several benefits contributed to making the DFIG the most extensively used generator in fluctuating speed wind power generation, outperforming any other architecture. Examples of such benefits include the capacity to manage power with a partial-sized converter in the rotor, lowering power losses and costs, minimizing efforts on mechanical parts, noise minimization, active and reactive power control, and an adjustable power factor [4, 5]. As illustrated in Fig. 1, the DFIG stator is directly related to the grid, and two bidirectional converters link the rotor to the transmission line through a DC-link voltage. The rotor side converter (RSC) is used to regulate the unit power factor and manage active and reactive power transferred with the grid. Instead, the grid side converter (GSC) is used to keep the DC Link voltage stable [5]. A multivariable, nonlinear mathematical model characterizes the DFIG. The magnetic flux and electromagnetic torque are intrinsically tied. As a result, DFIG control is more complicated and challenging than DC machine control [6]. Several strategies for controlling the DFIG have been presented. Field-Oriented Control (FOC) is widely utilized in WECS based on the wound rotor induction

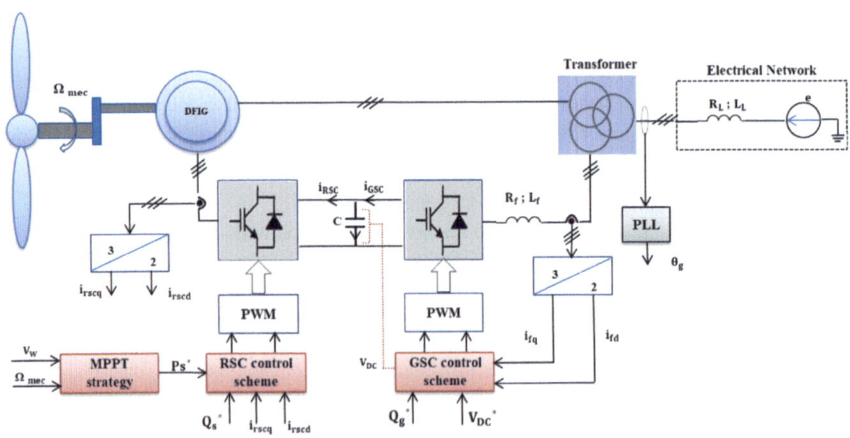

Fig. 1 Wind Power Plant System (WPPS) scheme

generator and has demonstrated better efficiency. The vector control is the simplest command scheme, and it is built with Proportional-Integral (PI) action controllers. Furthermore, this vector control allows the DFIG to be controlled as an independently excited DC machine, making it highly attractive in the field [6]. The typical field-oriented control achieves separated regulation of active and reactive power by controlling the d- and q-axis rotor currents in the reference frame aligned with the stator flux [7]. Besides, the Direct Vector Control (DVC) is a field-oriented control variant that is the simplest because it requires just two regulators to manage the powers by manipulating the rotor voltages directly. While the other variant is the Indirect Vector Control (IVC). There are two types of this variant. The first employs two regulators to control the power in an open loop. The second, on the other hand, regulates the powers in the closed-loop. The latter is more complicated and necessitates the use of four regulators to govern the stator powers and rotor currents [8]. Bouderbala et al. (2018) have suggested and compared the variants of the vector control based on PI controller adjusted by the compensation method for the wind power plant system [9]. However, one significant disadvantage of the PI regulator is that the parameters adjustment necessitates precise knowledge of system characteristics. As a result, each change in these parameters has an unfavorable impact on the performance of the control method [10].

Due to the nonlinear interaction between machine currents and rotor speed, and the existence of nonlinearity in the torque formula, nonlinear control is preferable in a DFIG-driven wind power plant system. Examples of nonlinear approaches are Linearization Feedback, Sliding Mode Control, and Backstepping. The First is a popular method for regulating nonlinear systems. The procedure includes developing a conversion of the nonlinear process into an analogous linear model using variable changes and a sufficient control input. Alrifai et al. (2016) have applied the Feedback Linearization Controller to pilot and control the powers separately of the wind energy system [11]. However, Conventional feedback linearization controllers are sensitive to generator settings changes and disturbances, necessitating new control strategies for robust and optimal performance in the face of nonlinear system dynamics [12]. The second is a nonlinear control approach that modifies the dynamics of a nonlinear system by utilizing a discontinuous control signal that drives the system to "slide" over a cross-section of the system's normal behavior [13]. The importance of sliding mode control derives from its high precision, fast dynamic response, stability, and robustness against uncertainties/variations of internal parameters and external disturbances [14]. However, the major drawbacks of the Sliding Mode Control (SMC) are the chattering phenomenon, which is the significant oscillations observed in the simulation results, and sensor drift is caused by high-frequency disturbance. The third is a nonlinear method for developing stabilizing controllers for a group of nonlinear dynamical systems. The latter is comprised of sub-systems that spread out from an infinitely complex subsystem that can be controlled in another manner. Because the system is recurrent, the researcher may begin the configuration with the known-stable system and "back out" novel controllers that gradually stabilize each anterior subsystem. Once the last external rule is achieved, the procedure finishes [15]. Backstepping control has several advantages, including assured global and regional stability, a focus on robustness, and computable transient performance. Some of the Backstepping disadvantages are the information regarding all system states is required and the presence of steady-state error, which can be judged acceptable and can be eliminated by proposing the Integrator Backstepping as a variant of Backstepping as cited in [16].

On the other hand, the maximum generated power from wind energy is determined by the turbine's characteristics and the wind's fluctuating speed. Maximum Power Point (MPP) is the name given to the method allowing getting the highest power. Several control strategies are created to implement the MPP algorithm. The MPP with optimal Tip Speed Ratio (TSR), the MPP with Optimal Torque Control (OTC), the MPP with Perturbation and Observation (PO), and the MPP with Power Signal Feed-back (PSF) are examples of these schemes [17, 18]. In this chapter, the TSR algorithm is used. However, this strategy has two variants the MPP with regulation speed which is based on a PI controller, and the MPP without regulation speed. In the first, the generator speed is managed to maintain an optimal TSR. The wind speed and rotor speed of the wind turbines are measured to apply this algorithm. In addition, the maximum value of the power coefficient and the optimal TSR are requested [19]. In the second, the torque control technique optimizes the wind generation system's torque, which is correlated with the best TSR [20]. This strategy necessitates the turbine angular speed measurement, the turbine mechanical torque value, the prior knowledge of the optimum torque, and maximum power coefficient; nevertheless, wind speed measurement is not required [21].

The MATLAB/Simulink environment is employed to model and perform a 5 MW DFIG. The latter is powered by a variable speed wind turbine in this chapter. A typical pulse width modulation controls the RSC and GSC (PWM). The main contribution of our chapter is the application and establishment of a control law for regulating a wind energy conversion system based on a varied wind profile speed. Therefore, this chapter is organized as the following:

- Wind power plant modeling and the MPP strategy are presented in Sect. 2.
- Section 3 proposes using the indirect vector control technique to manage the powers.
- Section 4 recommends the nonlinear Backstepping approach for the entire wind chain, including the MPP algorithm to enhance system performance.
- The outcomes of the simulation are explained and interpreted in Sect. 5.
- In the end, Sect. 6 presents the conclusion.

2 Wind Power Plant System Components Modeling

A wind turbine, a DFIG, a Pulse Width Modulation (PWM) AC/DC converter, and a PWM DC/AC converter constitute the wind power plant system. The synchronous d-q reference frame technique is employed to regulate the rotor side and the grid side power converter. The integrated Phase-Locked loop (PLL) technology synchronizes the GSC with the grid. A filter comprises a resistance (R_f), and inductance (L_f)

is used to alter the high-frequency oscillations. The GSC's filtered output is injected into the power system via a step-up transformer, as depicted in Fig. 1.

2.1 Wind Turbine (WT) Modeling

Using the cited below formula, the mechanical power on the turbine shaft (P_{Tu}) is determined using the specified variable wind speed V_w practiced to the WT blades [22]:

$$P_{Tu} = \frac{1}{2} \cdot C_P(\lambda, \beta) \cdot \rho_{air} \cdot \pi R_{blade}{}^2 V_w{}^3 \tag{1}$$

The wind turbine mechanical torque T_{Tu} $T_{Tu}T_{Tu}$ is expressed as below:

$$T_{Tu} = \frac{C_P(\lambda, \beta) \cdot \rho_{air} \cdot \pi \cdot R_{blade}{}^2 \cdot V_w{}^3}{2 \cdot \Omega_{Tu}} \tag{2}$$

where ρ_{air} is the density of the air (kg/m^3), R_{blade} is the radius of the WT blade (m), Ω_{Tu} is the rotational speed of the turbine, and C_P denotes the efficiency coefficient of the WT. C_P can be expressed as below [9, 16]:

$$C_P(\beta, \lambda) = [0.5 - 0.0167 \cdot (\beta - 2)] \cdot \sin\left(\frac{\pi(\lambda + 0.1)}{18.5 - 0.3 \cdot (\beta - 2)}\right) - 0.00184 \cdot (\lambda - 3) \cdot (\beta - 2) \tag{3}$$

This factor can be computed employing the tip speed (λ) and the blade pitch angle. The latter is set to 2° in order to have C_{pmax}. The expression of λ is defined by Eq. (4) [9]:

$$\lambda = \frac{R_{blade} \cdot \Omega_{Tu}}{V_w} \tag{4}$$

Keeping in mind that the total dynamics are carried back to the turbine rotor, the wind turbine mechanical equation is written as [9, 23]:

$$J_{tot} \cdot \frac{d\Omega_{mec}}{dt} + f_r \cdot \Omega_{mec} = T_{gb} - T_{el-m} \tag{5}$$

$$\text{Where} T_{gb} = \frac{T_{Tu}}{G_B} \text{and} G_B = \frac{\Omega_{mec}}{\Omega_{Tu}} \tag{6}$$

where J_{tot} is the global inertial of the wind turbine equipment, T_{Tu} is the WT torque, T_{elm} is the generator's electro-magnetic torque, f_r is the entire viscosity factor of friction, Ω_{mec} is the angular velocity of the gearbox's rotor (rad s^{-1}), and G_B is the gearbox factor.

2.2 *DFIG Modeling*

We applied the Park transformation to design the DFIG modeling by considering the simplifying hypotheses. In the d-q reference, the mathematical model of a DFIG is described as [19]:

$$\begin{cases} V_{sd} = R_s i_{sd} + \frac{d\varphi_{sd}}{dt} - \omega_s \varphi_{sq} \\ V_{sq} = R_s i_{sq} + \frac{d\varphi_{sq}}{dt} + \omega_s \varphi_{sd} \\ V_{rd} = R_r i_{rd} + \frac{d\varphi_{rd}}{dt} - \omega_r \varphi_{rq} \\ V_{rq} = R_r i_{rq} + \frac{d\varphi_{rq}}{dt} + \omega_r \varphi_{rd} \end{cases} \begin{cases} \varphi_{sd} = L_s.i_{sd} + M.i_{rd} \\ \varphi_{sq} = L_s.i_{sq} + M.i_{rq} \\ \varphi_{rd} = L_r.i_{rd} + M.i_{sd} \\ \varphi_{rq} = L_r.i_{rq} + M.i_{sq} \end{cases} \tag{7}$$

where V_r and V_s are volts, i_s and i_r are currents, φ_s and φ_r are fluxes, R_s and R_r are resistors of the windings, ω_s and ω_r are rotational speeds, Ls and L_r stand for coil inductances, and M represents mutual inductance. The letters r and s represent the rotor and stator, accordingly. The electromagnetic torque may be represented in the following way [19, 20]:

$$T_{em} = -P_p \cdot \frac{M}{L_s}\left(i_{rq}\varphi_{sd} - i_{rq}\varphi_{sd}\right) \tag{8}$$

The ratio of generator pole pairs is indicated by p_p.

2.3 *MPP Strategy*

To get the maximum power from the wind, we must constantly modify the turbine's rotor velocity in response to changes in wind speed [10]. The basic idea behind this instruction is to have a turbine angular velocity that permits for an appropriate speed ratio [20]. Figure 2 shows the MPP approach is created with speed regulation. This strategy means keeping the generator speed constant at its desired value, which is highest when the C_p is optimum. The electromagnetic torque generated by DFIG equals the desired value computed by the control, which is described as [24]:

$$T_{elm} = T_{elm-opt} \tag{9}$$

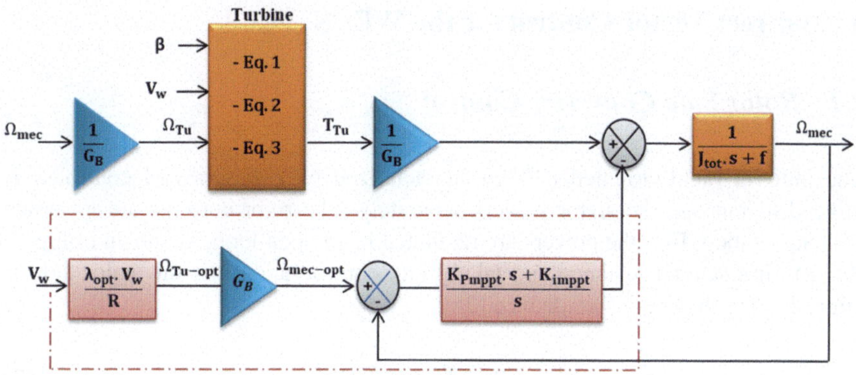

Fig. 2 MPP strategy with regulation speed based on PI controller

The optimum electromagnetic torque $T_{elm\text{-}opt}$ for achieving the optimal speed is provided as follows:

$$T_{elm-opt} = \left[K_{pmppt} + K_{imppt} \cdot \frac{1}{S}\right] \cdot [\Omega_{mec-opt} - \Omega_{mec}] \tag{10}$$

K_{pmppt} and K_{imppt} denote the PI regulator parameters. The optimal speed ($\Omega_{mec\text{-}opt}$) is:

$$\Omega_{mec-opt} = G_B \cdot \Omega_{Tu-opt}; \text{ With}\Omega_{Tu-opt} = \frac{V_w \cdot \lambda_{opt}}{R_{blade}} \tag{11}$$

Computation of the PI parameters for MPP. The pole compensation approach determines the PI controller settings. The system's time constant (T_{sys}) is:

$$T_{sys} = \frac{J_{tot}}{f_r} \tag{12}$$

The settings of the regulator are calculated as:

$$K_{imppt} = \frac{1}{\tau \cdot f_r} \text{ And } K_{pmppt} = \frac{-K_{imppt} \cdot J_{tot}}{f_r}; \text{ with}\tau = \frac{T_{sys}}{1000} \tag{13}$$

3 Indirect Vector Control for the WECS

3.1 Rotor Side Converter Control

The Indirect Field-Oriented (IFOC) characterized by two series PI controller is utilized to manage the stator powers separately. The rotor currents are governed in a closed loop. But, the powers are regulated in an open-loop, as shown in Fig. 3. By ignoring stator resistance and taking into consideration that the stator field is aligned along the d-axis, we deduce [25]:

$$\varphi_{sd} = \varphi_s, \varphi_{sq} = 0 \tag{14}$$

$$V_{sd} = 0; V_{sq} = V_s = \omega_s.\varphi_s$$

The rotor currents i^*_{rq} and i^*_{rd} are computed as [25]:

$$i^*_{rq} = -\frac{L_s}{M \cdot V_s} \cdot P^*_s \tag{15}$$

$$i^*_{rd} = -\frac{L_s}{M.V_{sq}} \cdot (Q^*_s - \frac{V_s^2}{\omega_s \cdot L_s}) \tag{16}$$

P^*_s represents the maximum generated power by the turbine as determined by the MPP method ($P^*_s = P^*_{Tu}$). For a null transfer of reactive power, the desired reactive power ($Q_s{}^*$) is set to zero. The voltage references are written down as follows:

$$V^*_{rq} = \left[i^*_{rq} - i_{rq}\right] \cdot \left[K_{prsc1} + K_{irsc1} \cdot \frac{1}{S}\right] + e_{rd} + Vs' \tag{17}$$

$$V^*_{rd} = \left[i^*_{rd} - i_{rd}\right] \cdot [K_{prsc2} + K_{irsc2} \cdot \frac{1}{S}] + e_{rq} \tag{18}$$

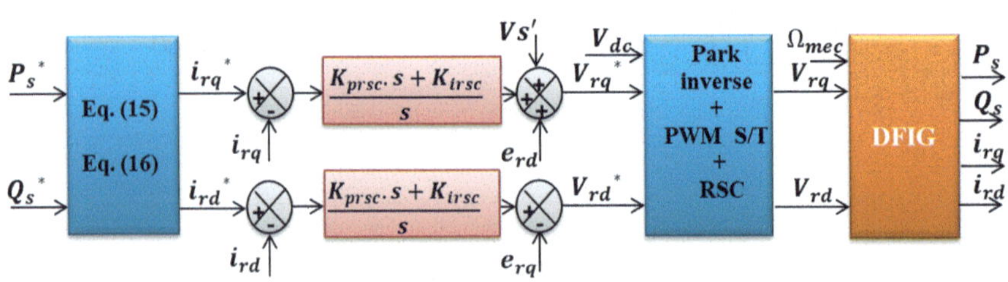

Fig. 3 Control scheme for the RSC

$$\text{Where } e_{rd} = g \cdot \omega_s \cdot \left(L_r - \frac{M^2}{L_s}\right) \cdot i_{rd};$$

$$e_{rq} = g \cdot \omega_s \cdot \left(L_r - \frac{M^2}{L_s}\right) \cdot i_{rq}\, Vs' = s \cdot \frac{V_s M}{L_s} \tag{19}$$

The expression of the response time is:

$$T_r = (L_r - \frac{M^2}{L_s})/R_r \tag{20}$$

Pole compensation is used to determine the PI settings, which are as follows:

$$K_{prsc} = \frac{1}{T_{rsc}}.(L_r - \frac{M^2}{L_s}) \text{ and } K_{irsc} = \frac{K_{prsc}.\, R_r}{(L_r - \frac{M^2}{L_s})}; \text{ with } T_{rsc} = \frac{T_r}{100} \tag{21}$$

3.2 Grid Side Power Converter Control

The GS Converter regulates DC-link voltage and reactive power transferred with the power system. The grid voltage is supposed to be aligned with the q-axis. Figure 4 depicts the GSC's control scheme. As a result, the grid voltages are stated as follows:

$$V_{gd} = 0 \text{ and } V_{gq} = V_g \tag{22}$$

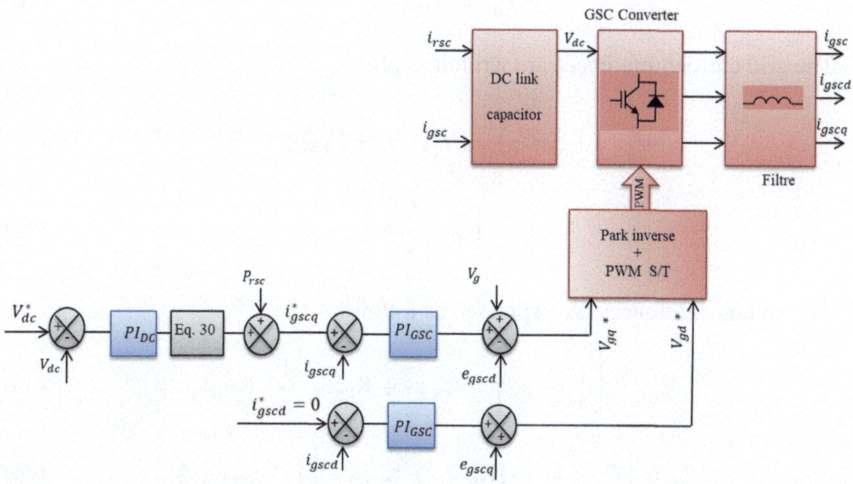

Fig. 4 Scheme control of the GSC

The control is accomplished by adjusting the current reference frame. As a result, the voltages and powers may be expressed as [26]:

$$V_{gscd} = -[R_f + L_f.s] \cdot i_{gd} + \omega_s \cdot L_f \cdot i_{gq} \quad (23)$$

$$V_{gscq} = -[R_f + L_f \cdot s] \cdot i_{gq} - \omega_s \cdot L_f \cdot i_{gd} + V_g \quad (24)$$

$$P_g = V_g \cdot i_{gq} \quad (25)$$

$$Q_g = V_g \cdot i_{gd} \quad (26)$$

The relationship between the output powers of the converters is as follows [26]:

$$V_{dC} \cdot i_C = P_{GSC} - P_{RSC} \quad (27)$$

The power of the GSC (P_g) is stated as follows:

$$P_g = P_{GSC} = V_{dC} \cdot i_C + P_{RSC} \quad (28)$$

The power of RSC is P_{RSC}, which can be defined by:

$$P_{RSC} = V_{DC} \cdot\ i_{RSC} \quad (29)$$

Therefore, the DC-link power (P^*_{dc}) is described as below:

$$P^*_{dc} = V_{DC} \cdot\ i^*_c. \quad (30)$$

The grid current references are written as [26]:

$$i^*_{gq} = \frac{1}{V_g} \cdot \left(V^*_{DC} \cdot i^*_c + P_{RSC}\right) \quad (31)$$

$$i^*_{gd} = \frac{Q^*_g}{V_g} \quad (32)$$

The voltage references are expressed as follows:

$$V^*_{gd} = \left[i^*_{gd} - i_{gd}\right] \cdot [K_{pgsc2} + K_{igsc2} \cdot \frac{1}{S}] + e_{gscq} \quad (33)$$

$$V^*_{gq} = \left[i^*_{gq} - i_{gq}\right] \cdot [K_{pgsc1} + K_{igsc1} \cdot \frac{1}{S}] - e_{gscd} + V_g \quad (34)$$

$$\text{Wheree}_{gscq} = \omega_s \cdot L_f \cdot i_{gq};\ e_{gscd} = \omega_s \cdot L_f \cdot i_{gd} \quad (35)$$

The grid currents is given as follows [10, 26]:

$$i_{gq} = \frac{-1}{[R_f + L_f.s]} \cdot \left(V_{gq}^* - \omega_s \cdot L_f \cdot i_{gd} - V_{DC} \cdot S_q\right) \tag{36}$$

$$i_{gd} = \frac{-1}{[R_f + L_f.s]} \cdot \left(V_{gd}^* + \omega_s \cdot L_f \cdot i_{gq} - V_{DC} \cdot S_d\right) \tag{37}$$

In the d-q referee, the GSC's switching modes are Sd and Sq. As illustrated in Fig. 4, the PI_{DC} controller is utilized to regulate the DC-link voltage at its reference. Therefore, The PI $_{DC}$ regulator parameters (K $_{pDC}$, K $_{iDC}$) are calculated in this way [10]:

$$K_{pDC} = 2 \cdot \xi \cdot \omega \cdot c; K_{iDC} = \omega^2 \cdot c \tag{38}$$

ξ represents the damping factor.

The currents i_{gq} and i_{gd} that pass through the RL filter are controlled by using a PI_{GSC} regulator. As a result, the controlled system's temporal constant is stated as:

$$T_s = \frac{L_f}{R_f} \tag{39}$$

The PI controller gains are given as:

$$K_{pgsc} = \frac{L_f}{T}; \; K_{igsc} = \frac{R_f}{T}; \text{ with } T = \frac{T_s}{10} \tag{40}$$

4 Backstepping Control Applied for the WECS

4.1 Backstepping Overview

The Backstepping methodology is a systematic and recursive method of generating nonlinear control rules from Lyapunov functions that ensures the stability of each synthesis step by step [25, 27]. A virtual control is produced at each phase of the process to guarantee the system's convergence towards its equilibrium state. This approach enables the synthesis of control rules while accounting for potential disruptions or a lack of knowledge of system characteristics. The essential principle behind Backstepping control is to make looping systems equal to sub-systems of order one in a stable cascade in the sense of Lyapunov, which provides it resilience and global asymptotic stability [15].

4.2 Application of Backstepping for MPP

According to the expression of the dynamical speed of rotation of the wound rotor induction generator given by Eq. (5), the Backstepping technique is utilized to develop the control. So, the error variable is defined, in the first step, as follows:

$$\varepsilon_\Omega = \Omega^*_{mec} - \Omega_{mec} \quad (41)$$

Equation (42) provides the DFIG rotation speed derivative:

$$\frac{d\Omega_{mec}}{dt} = \frac{1}{J_{tot}} \cdot (T_g - T_{elm} - f_r \cdot \Omega_{mec}) \quad (42)$$

Therefore, the error derivative is:

$$\dot{\varepsilon}_\Omega = \dot{\Omega}^*_{mec} - \dot{\Omega}_{mec} \quad (43)$$

By replacing Eq. (42) in Eq. (43) we obtain:

$$\dot{\varepsilon}_\Omega = \dot{\Omega}_{mec} - \frac{1}{J_{tot}} \cdot (T_g - T_{elm} - f_r \cdot \Omega_{mec}) \quad (44)$$

Choosing L_f as a Lyapunov function [19, 28]:

$$L_{f-mppt} = \frac{1}{2}\varepsilon^2_\Omega \quad (45)$$

Computing the derivative of L_f as:

$$\dot{L}_{f-mppt} = \varepsilon_\Omega \cdot \dot{\varepsilon}_\Omega = \varepsilon_\Omega \cdot (\dot{\Omega}^*_{mec} - \dot{\Omega}_{mec}) \quad (46)$$

Equation (46) can be written as:

$$\dot{L}_f = \varepsilon_\Omega \cdot \left[\dot{\Omega}^*_{mec} - \frac{1}{J_{tot}} \cdot \left(T_g - T_{elm} - f_r \cdot \Omega_{mec}\right)\right] \quad (47)$$

In order to examine the system's stability, we develop the Stabilizing function of Lyapunov, often known as virtual control, as [27, 28]:

$$\dot{L}_f = -C_{mppt}.\varepsilon^2_\Omega \quad (48)$$

In the Eq. (48), C_{mppt} defines a positive number. T_{elm} is a virtual function that can be described as follows:

$$T^*_{elm} = -J_{tot}\left(\dot{\Omega}^*_{mec} + C_{mppt} \cdot \varepsilon_\Omega\right) + T_g - f_r \cdot \Omega_{mec} \quad (49)$$

4.3 Application of Backstepping for RSC

The rotor currents will be used to build the virtual command. The following is the derivation of the rotor current mathematical formulas:

$$\begin{cases} \frac{di_{rd}}{dt} = \frac{1}{a}(V_{rd} - R_r i_{rd} + b \cdot i_{rq}) \\ \frac{di_{rq}}{dt} = \frac{1}{a}(V_{rq} - R_r i_{rq} - b \cdot i_{rd} - c) \end{cases} \tag{50}$$

where, $a = L_r \cdot \left(1 - \frac{M^2}{L_r \cdot L_s}\right)$; $b = g \cdot \omega_s L_r \left(1 - \frac{M^2}{L_r \cdot L_s}\right)$; $c = \frac{M \cdot V_s \cdot g}{L_s}$

We compute the derivation of the rotor current errors as follows:

$$\frac{d\varepsilon_{rsc1}}{dt} = \frac{di_{rq}^*}{dt} - \frac{di_{rq}}{dt} \text{ And } \frac{d\varepsilon_{rsc2}}{dt} = \frac{di_{rd}^*}{dt} - \frac{di_{rd}}{dt} \tag{51}$$

Substituting the derivative of currents with respect to time into Eq. (51) generates:

$$\frac{d\varepsilon_{rsc1}}{dt} = \frac{di_{rq}^*}{dt} - \frac{1}{a}(V_{rq} - R_r i_{rq} - b \cdot i_{rd} - c) \tag{52}$$

$$\frac{d\varepsilon_{rsc2}}{dt} = \frac{di_{rd}^*}{dt} - \frac{1}{a}(V_{rd} - R_r i_{rd} + b \cdot i_{rq}) \tag{53}$$

In a second phase, the Lyapunov candidate function is formulated as having:

$$L_{f-rsc} = \frac{\varepsilon_{rsc1}^2}{2} + \frac{\varepsilon_{rsc2}^2}{2} \tag{54}$$

The Lyapunov function is derived in this way:

$$\dot{L}_{f-rsc} = \varepsilon_{rsc1} \cdot \dot{\varepsilon}_{rsc1} + \varepsilon_{rsc2} \cdot \dot{\varepsilon}_{rsc2} \tag{55}$$

Substituting the errors derivation in Eq. (55), the Lyapunov derivative can be given by:

$$\begin{aligned} \dot{L}_{f-rsc} = {} & \varepsilon_{rsc1} \cdot \left[\frac{di_{rq}^*}{dt} - \frac{1}{a}(V_{rq} - R_r i_{rq} - b \cdot i_{rd} - c)\right] \\ & + \varepsilon_{rsc2} \cdot \left[\frac{di_{rd}^*}{dt} - \frac{1}{a}(V_{rd} - R_r i_{rd} + b \cdot i_{rq})\right] \end{aligned} \tag{56}$$

According to the Lyapunov function, the derivation of L_{f-rsc} must be negative to maintain system stability, and is shown below:

$$\dot{L}_{f-rsc} = -C_{rsc1} \cdot \varepsilon_{rsc1}^2 - C_{rsc2} \cdot \varepsilon_{rsc2}^2 \tag{57}$$

where C_{rsc1} and C_{rsc2} are positive constant. By equalizing the two Eqs. (56) and (57), we obtain:

$$\frac{di_{rq}^*}{dt} - \frac{1}{a}(V_{rq} - R_r i_{rq} - b \cdot i_{rd} - c) = -C_{rsc1} \cdot \varepsilon_{rsc1} \quad (58)$$

$$\frac{di_{rd}^*}{dt} - \frac{1}{a}(V_{rd} - R_r i_{rd} + b \cdot i_{rq}) = -C_{rsc2} \cdot \varepsilon_{rsc2} \quad (59)$$

Therefore, the virtual command V_{rd} and V_{rq} are deduced directly as follows:

$$V_{rq}^* = a \cdot \left[\frac{di_{rq}^*}{dt} + \frac{R_r}{a} i_{rq} + \frac{b}{a} \cdot i_{rd} + \frac{c}{a} + C_{rsc1} \cdot \varepsilon_{rsc1}\right] \quad (60)$$

$$V_{rd}^* = a \cdot \left[\frac{di_{rd}^*}{dt} + \frac{R_r}{a} i_{rd} - \frac{b}{a} \cdot i_{rq} + C_{rsc2} \cdot \varepsilon_{rsc2}\right] \quad (61)$$

4.4 *Application of Backstepping for GSC and DC Link Voltage*

The Backstepping process is established to maintain the DC link voltage invariant and generate the current desired value for controlling the powers. The error between the voltages of DC-link is expressed as:

$$\varepsilon_{dc} = V_{dc}^* - V_{dc} \quad (62)$$

The derivative of the error is given by:

$$\dot{\varepsilon}_{dc} = \dot{V}_{dc}^* - \dot{V}_{dc} \quad (63)$$

Choosing the first Lyapunov function associated with continuous bus voltage error:

$$L_{f-dc} = \frac{\varepsilon_{dc}^2}{2} \quad (64)$$

The derivative of Eq. (64) gives:

$$\dot{L}_{f-dc} = \varepsilon_{dc} \cdot \dot{\varepsilon}_{dc} = \varepsilon_{dc}(\dot{V}_{dc}^* - \frac{i_c}{c}) \quad (65)$$

The Eq. (65) can be modified to meet the requirement ($\dot{L}_{f-dc} < 0$) as follows:

$$\dot{L}_{f-dc} = -C_{dc} \cdot \varepsilon_{dc}^2 \tag{66}$$

where C_{dc} is a positive gain of the Backstepping. From Eqs. (65) and (66), we get:

$$\dot{V}_{dc}^* - \frac{i_c}{c} = -C_{dc} \cdot \varepsilon_{dc} \tag{67}$$

Equation (67) permits the synthesis of the continuous bus current desired (i_c^*), such as:

$$i_c^* = (\dot{V}_{dc}^* + C_{dc} \cdot \varepsilon_{dc}) \tag{68}$$

On the other hand, we have:

$$i_c^* = \frac{P_{dc}^*}{V_{dc}} \text{ And } i_{gq}^* = \frac{P_{gsc}^*}{V_g} \tag{69}$$

As a result, as a function of the DC-link voltage, the quadrature grid current reference may be stated as:

$$i_{gq}^* = \frac{C \cdot V_{dc} \cdot \left(\dot{V}_{dc}^* + C_{dc} \cdot \varepsilon_{dc}\right) + P_{rsc}}{V_g} \tag{70}$$

Grid current time derivative expressions are as described in the following:

$$\begin{gathered} \frac{di_{gd}}{dt} = \frac{1}{L_f}(-V_{gd} - R_f i_{gd} + \omega_s \cdot L_f \cdot i_{gq}) \\ \frac{di_{gq}}{dt} = \frac{1}{L_f}(-V_{gq} - R_f i_{gq} - \omega_s \cdot L_f \cdot i_{gd} + V_g) \end{gathered} \tag{71}$$

The grid current errors' derivative is provided by:

$$\frac{d\varepsilon_{gsc1}}{dt} = \frac{d\,i_{gq}^*}{dt} - \frac{di_{gq}}{dt} \text{ And } \frac{d\varepsilon_{gsc2}}{dt} = \frac{d\,i_{gd}^*}{dt} - \frac{d\,i_{gd}}{dt} \tag{72}$$

The errors derivative becomes as follows:

$$\frac{d\varepsilon_{gsc1}}{dt} = \frac{di_{gq}^*}{dt} - \frac{1}{L_f}(-V_{gq} - R_f i_{gq} - \omega_s \cdot L_f \cdot i_{gd} + V_g) \tag{73}$$

$$\frac{d\varepsilon_{gsc2}}{dt} = \frac{di_{gd}^*}{dt} - \frac{1}{L_f}(-V_{gd} - R_f i_{gd} + \omega_s \cdot L_f \cdot i_{gq}) \tag{74}$$

Lyapunov function is defined by:

$$L_{f-gsc} = \frac{\varepsilon_{gsc1}^2}{2} + \frac{\varepsilon_{gsc2}^2}{2} \tag{75}$$

The derivation of Lyapunov function is expressed by:

$$\dot{L}_{f-gsc} = \varepsilon_{gsc1} \cdot \dot{\varepsilon}_{gsc1} + \varepsilon_{gsc2} \cdot \dot{\varepsilon}_{gsc2} \tag{76}$$

$$\begin{aligned} \dot{L}_{f-gsc} = \varepsilon_{gsc1}\left[\frac{d\,i_{gq}^*}{dt} - \frac{1}{L_f}(-V_{gq} - R_f\,i_{gq} - \omega_s \cdot L_f \cdot i_{gd} + V_g)\right] \\ + \varepsilon_{rsc2} \cdot \left[\frac{d\,i_{gd}^*}{dt} - \frac{1}{L_f}(-V_{gd} - R_f\,i_{gd} + \omega_s L_f i_{gq})\right] \end{aligned} \tag{77}$$

According to Lyapunov, the derivation of L_{f-gsc} must be smaller than zero ($L_{f-gsc} < 0$) in order to preserve the system's stability:

$$\dot{L}_{f-rsc} = -C_{gsc1} \cdot \varepsilon_{gsc1}^2 - C_{gsc2} \cdot \varepsilon_{gsc2}^2 \tag{78}$$

The grid current control law can be determined as follows:

$$\frac{d\,i_{gq}^*}{dt} - \frac{1}{L_f}(-V_{gq} - R_f\,i_{gq} - \omega_s \cdot L_f \cdot i_{gd} + V_g) = -C_{gsc1} \cdot \varepsilon_{gsc1} \tag{79}$$

$$\frac{di_{gd}^*}{dt} - \frac{1}{L_f}(-V_{gd} - R_f i_{gd} - \omega_s \cdot L_f \cdot i_{gq}) = -C_{gsc2} \cdot \varepsilon_{gsc2} \tag{80}$$

As a conclusion, the formulations of the actual global commands V_{gd} and V_{gq} are described as follows:

$$V_{gq} = -L_f \cdot \left[\frac{di_{gq}^*}{dt} + C_{gsc1} \cdot \varepsilon_{gsc1}\right] - R_f i_{gq} - \omega_s \cdot L_f \cdot i_{gd} + V_g \tag{81}$$

$$V_{gd} = -L_f \cdot \left[\frac{d\,i_{gd}^*}{dt} + C_{gsc2} \cdot \varepsilon_{g2}\right] - R_f\,i_{gd} + \omega_s.L_f \cdot i_{gq} \tag{82}$$

5 Simulation Results

A power plant system is modeled using Matlab software and simulated under variable wind speed conditions to verify the effectiveness of the nonlinear Backstepping Controller. The settings of the wind energy system used in the simulation are given in Table 1. The wind profile in Fig. 5 begins at 9 m s^{-1}, immediately shifts to 11 m s^{-1} at 0.08 s, and then suddenly changes to 12.5 m s^{-1} at 0.2 s. The speed is then reduced to 10 m s^{-1} at 0.7 s to test the performance and excellence of the suggested technique. The reactive powers Q_g^* and Q_s^* are fixed to 0 VAR to keep a

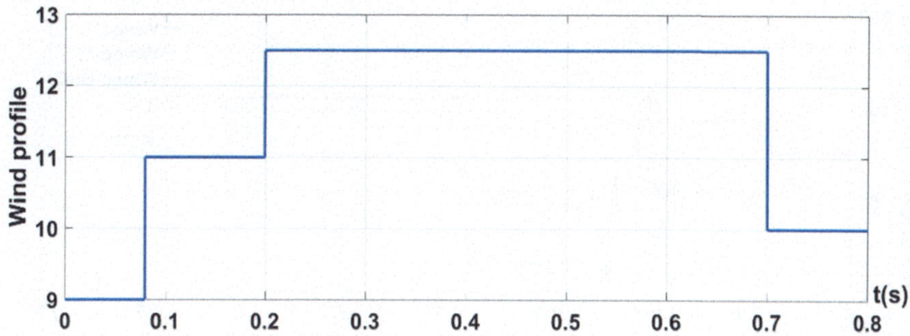

Fig. 5 Variable wind speed profile (m/s)

power factor at unity. The system is governed by two technologies that aim to illustrate the merits of the proposed technology by managing the power and maintaining a consistent DC-link voltage. The first option is to use ordinary PI controllers to establish an indirect control vector. The pole compensation approach is employed to evaluate the gains of the various PI regulators listed in Table 2. The nonlinear Backstepping Controller (BSC) is then utilized for maximum power point tracking and established for regulating wind energy system components. The characteristics of this methodology are evaluated depending on the results of repeated tests.

The nonlinear Backstepping and indirect field-oriented control associated with classical PI regulators results are achieved and compared. Figure 6 depicts the wind turbine efficiency calculated by the two techniques. As you can see, the nonlinear Backstepping regulates the C_p at its highest value, $C_{pmax} = 0.5$, regardless of wind speed change. On the other hand, Fig. 7 depicts the mechanical speed estimated by the standard PI controller and the Backstepping controller (BSC). The rotational speed is employed to operate the wound rotor induction generator. The latter is fed by the mechanical power generated. The proposed approach improves the temporal response with a negligible overshoot in the region of 3 rad/s. Figure 8 indicates that the stator active power follows its reference. The latter is the mechanical power produced by the wind turbine and obtained by the MPP strategy. It is clear that the

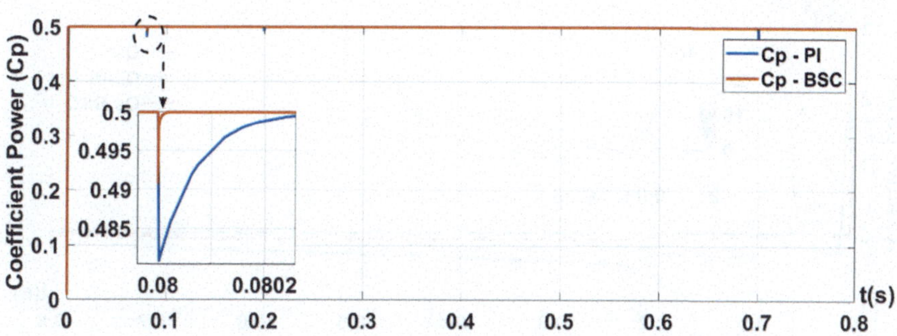

Fig. 6 Coefficient of power

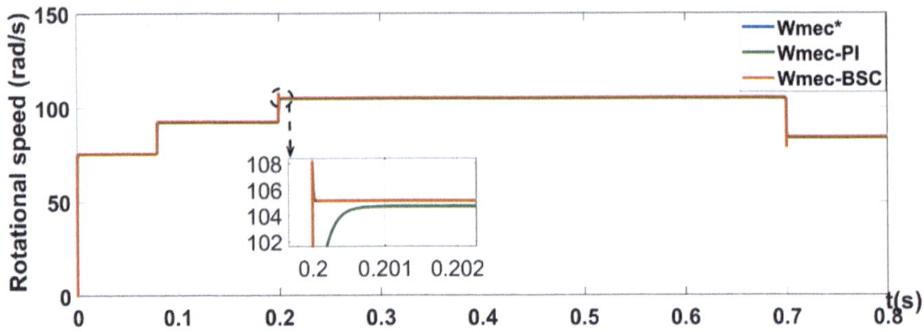

Fig. 7 Rotational speed (rad/s)

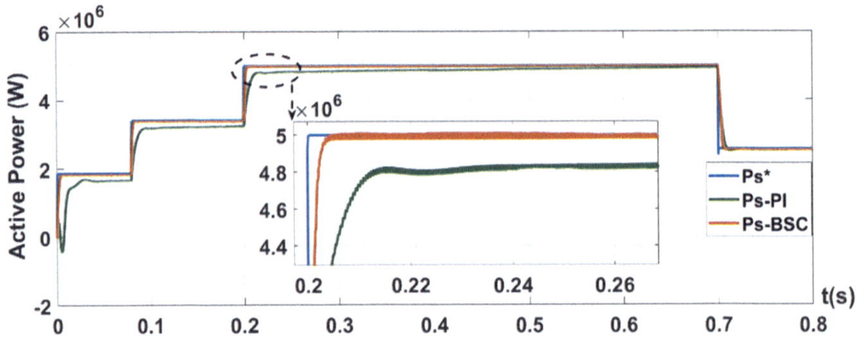

Fig. 8 Active power of the stator (W)

time response is improved from 10.4 (ms) to 1.9 (ms), and the static error is lowered from 2.64 to 0.16%. Weak ripples in active power, on the other hand, can be ignored. Figure 9 depicts the reactive power of the stator. The power coefficient is represented in Fig. 10. Figure 11 illustrates the voltage of the continuous bus system, which is calculated by using the standard PI Controller and the Backstepping Controller

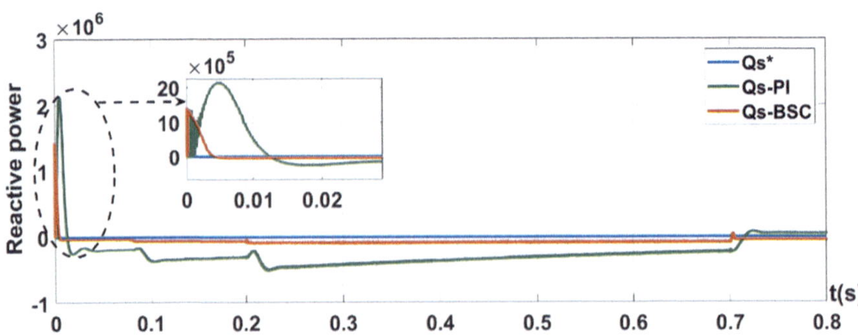

Fig. 9 Reactive power of the stator (VAR)

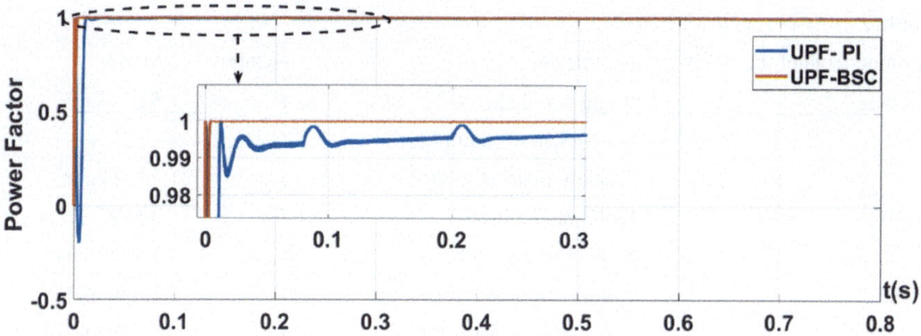

Fig. 10 Power coefficient

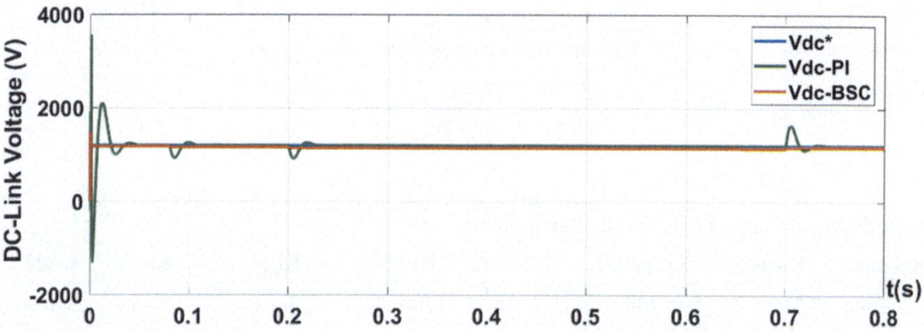

Fig. 11 DC-link voltage (V)

(BSC). Notably, the BSC approach minimizes the overshoot of the DC link voltage with a ratio of 2.38 and improves the response time.

According to the simulation findings, BSC has the fastest reaction time when opposed to standard PI. Nonlinear Backstepping demonstrates its durability against considerable variations in the speed reference, which considerably reduces the time response, specifically when the wind profile quickly changes.

6 Conclusion

The nonlinear Backstepping controller for governing the wind power plant system using a wound rotor induction generator is effectively developed in this chapter. The power grid is directly linked to the DFIG's stator, while it is connected to the rotor via AC/DC/AC power converters. Under fluctuating wind speed circumstances, the WECS model is simulated by employing Indirect Field-Oriented Control (IFOC) based on ordinary PI regulators. The nonlinear Backstepping Controller is then constructed. The Lyapunov function determines the Backstepping procedure, which guarantees system stability. When compared to indirect field-oriented Control,

Table 1 A list of input variables

Components of the WPP	Parameters	Symbol	Value
Turbine	Radius of blade	R	51,583 m
	Coefficient of multiplier	G_B	47.23
	Total moment of inertia	J_{tot}	1000 kg m^2
DFIG	DFIG rated power	Ps	5 MW
	Stator leakage inductance	Ls	1.2721 mH
	Rotor resistance	Rr	1.446 mΩ
	Rotor leakage inductance	Lr	1.1194 mH
	Mutual inductance	M	0.55187 mH
	Rated stator line to line voltage	Vs	950 V
Capacity	DC-link capacitance	C	4400 μF
	DC link Voltage reference	V_{dc*}	1200 V
Filter RL	Resistor of the filter	R_f	20 Ω
	Inductance of the filter	L_f	0.08 H

Table 2 Vector control gains of the PI regulators

Kp-mppt	Ki-mppt	Kp-rsc1/2	Ki-rsc1/2	Kp-gsc1/2	Ki-gsc1/2	Kp-dc	Ki-dc
−7.2e+6	51.87	0.1446	0.2376	200	5e+4	1.48	396

simulation outcomes affirm that the suggested strategy delivers a quicker transient response with reduced tracking error, excellent stability, and good robustness.

Appendix

See Tables 1 and 2.

References

1. Mensou, S., Essadki, A., Minka, I., Nasser, T., Bououlid Idrissi, B.: Backstepping controller for a variable wind speed energy conversion system based on a DFIG. Int. J. Electr. Comput. Eng. **12**(9), 598–604 (2018)
2. Ouassaid, M., Elyaalaoui, K., Cherkaoui, M.: Sliding mode control of induction generator wind turbine connected to the grid. In: Vaidyanathan S., Volos C. (eds.) Advances and Applications in Nonlinear Control Systems. Studies in Computational Intelligence, vol. 635. Springer, Cham (2016)
3. Sawant, M., Thakare, S., Rao, A.P., Feijóo-Lorenzo, A.E., Bokde, N.D.: A review on state-of-the-art reviews in wind-turbine- and wind-farm related topics. Energies **14** (2021)

4. Chetouani, E., Errami, Y., Obbadi, A., Sahnoun, S.: Maximum power point tracking design using particle swarm optimization algorithm for wind energy conversion system connected to the grid. Renew. Energy Syst. 445–470 (2021)
5. Anca, D.H.: Wind turbine technologies. In: Letcher, T.M. (eds.) Wind Energy Engineering. Academic Press, pp. 145–160 (2017)
6. Ihedrane, Y., El Bekkali, C., Bossoufi, B., Bouderbala, M.: Control of power of a DFIG generator with MPPT technique for wind turbines variable speed. In: Derbel, N., Zhu, Q. (eds.) Modeling, Identification and Control Methods in Renewable Energy Systems. Green Energy and Technology. Springer, Singapore, pp. 105–129 (2019)
7. Jose, J.T., Chattopadhyay, A.B: Mathematical formulation of feedback linearizing control of doubly fed induction generator including magnetic saturation effects. Math. Probl. Eng. **2020**, 1–10 (2020)
8. Mazouz, F., Belkacem, S., Harbouche, Y., Abdessemed, R., Ouchen, S.: Active and reactive power control of a DFIG for variable speed wind energy conversion. In: Proceedings of the 6th International Conference on Systems and Control, University of Batna 2, Batna, Algeria (2017)
9. Bouderbala, M., Bossoufi, B., Lagrioui, A., Taoussi, M., Aroussi, H., Ihedrane, Y.: Direct and indirect vector control of a doubly fed induction generator based in a wind energy conversion system. Int. J. Electr. Comput. Eng. (IJECE) **9**(3), 1531–1540 (2018)
10. Chetouani, E., Errami, Y., Obbadi, A., Sahnoun, S.: Optimal tuning of PI controllers using adaptive particle swarm optimization for doubly-fed induction generator connected to the grid during a voltage dip. Bull. Electr. Eng. Inform. **10**(5), 1–12 (2021)
11. Alrifai, M., Zribi, M., Rayan, M.: Feedback linearization controller for a wind energy power system. Energies **9**, 771 (2016)
12. Amin, I.K., Uddin, M.N.: Nonlinear control operation of DFIG-based WECS incorporated with machine loss reduction scheme. IEEE Trans. Power Electron. **35**(7), 7031–7044 (2020)
13. Solis, C., Clempner, J., Poznyak, A.: Robust extremum seeking for a second order uncertain plant using a sliding mode controller. Int. J. Appl. Math. Comput. Sci. **29**(4), 703–712 (2020)
14. Pimpale, Y., Parvat, B.J.: Design of sliding mode control for nonlinear uncertain system. Int. J. Adv. Res. Innov. Ideas Educ. **4**(5), 331–336 (2018)
15. Vaidyanathan, S., Azar, A.T.: An introduction to backstepping control. Backstepping Control of Nonlinear Dynamical Systems. Elsevier Inc., pp. 1–32 (2021)
16. Chetouani, E., Errami, Y., Obbadi, A., Sahnoun, S.: Hybrid control using adaptive particle swarm optimization and integral backstepping control of grid-connected doubly fed induction generator. Trends Sci. (TIS) **18**(23), 712–729 (2021)
17. Chavero-Navarrete, E., Trejo-Perea, M., Jáuregui-Correa, J.C., Carrillo-Serrano, R.V., Ríos-Moreno, J.G.: Expert control systems for maximum power point tracking in a wind turbine with PMSG: state of the art. Appl. Sci. **9**(12), 24–69 (2019)
18. Zhang, Y., Zhang, L., Liu, Y.: Implementation of maximum power point tracking based on variable speed forecasting for wind energy systems. Processes **7**(3), 158–176 (2019)
19. Chetouani, E., Errami, Y., Obbadi, A., Sahnoun, S.: Design of optimal backstepping control for a wind power plant system using the adaptive weighted particle swarm optimization. Int. J. Intell. Eng. Syst. **14**(6), 125–136 (2021)
20. Chojaa, H., Derouich, A., Taoussi, M., Zamzoum, O., Hanafi, A.: An improved performance variable speed wind turbine driving a doubly fed induction generator using sliding mode strategy. In: IEEE 2nd International Conference on Electronics, Control, Optimization and Computer Science. IEEE, Kenitra, Morocco, pp. 1–8 (2020)
21. Saavedra-Montes, A.J., Ramos-Paja, C.A., Ramírez-Gómez, C.A.: Model-based maximum power point tracking for wind generators. Revista Facultad de Ingeniería **79**, 75–83 (2016)
22. Errami, Y., Obbadi, A., Sahnoun, S.: An improved control of grid integrated doubly fed induction generator. Int. J. Power Energy Convers. **12**(3), 219–235 (2021)
23. Errami, Y., Obbadi, A., Sahnoun, S.: Control of PMSG wind electrical system in network context and during the MPP tracking process. Int. J. Syst. Control Commun. **11**(2), 200–225 (2020)

24. Dahbi, A., Nait-Said, N., Nait-Said, M.: A novel combined MPPT-pitch angle control for wide range variable speed wind turbine based on neural network. Int. J. Hydrogen Energy **41**(22), 9427–9442 (2016)
25. Chetouani, E., Errami, Y., Obbadi, A., Sahnoun, S.: Backstepping and indirect vector control for rotor side converter of doubly fed-induction generator with maximum power point tracking. In: Motahhir, S., Bossoufi, B. (eds.) Digital Technologies and Applications. ICDTA 2021. Lecture Notes in Networks and Systems, vol. 211, Springer, Cham, pp. 1711–1723 (2021)
26. Elazzaoui, M.: Modeling and control of a wind system based doubly fed induction generator: optimization of the power produced. J. Electr. Electron. Syst. **4**(1) (2015)
27. Chetouani, E., Errami, Y., Obbadi, A., Sahnoun, S., Wadawa, B.: Nonlinear backstepping with integral action for wind power plant based on doubly fed induction generator connected to the non-ideal grid. Technol. Econ. Smart Grids Sustain. Energy **7**(4) (2022)
28. Errami, Y., Obbadi, A., Sahnoun, S., Benhmida M., Ouassaid, M., Maaroufi, M.: Design of a nonlinear backstepping control strategy of grid interconnected wind power system based PMSG. AIP Conf. Proc. **1758** (2016)

Improving LVRT Capability of a Wind Turbine During a Voltage Sag in the Electrical Network

Sara Mensou, Zineb Mekrini, Ahmed Essadki, Tamou Nasser, and Badr Bououlid Idrissi

Abstract The wind turbine (WT) based on a Doubly Fed Induction Generator (DFIG) has experienced a fast evolution in the world, which leads to an increasing insertion of this type of generators in WTs connected to the electrical networks. When a voltage dips appears on the grid, the aero-generator can lose its operation, the voltage drop produces high peaks currents on the generator winding, without protection, the power converters will suffer also from over-current limit, as a consequence, it may be damaged and the DFIG be destroyed. In this chapter a new Direct Power Control (DPC) approach was applied, to improve the aero-generator operation under the network faults. The DPC approach was designed using the nonlinear Back-Stepping Controller (BSC) to ensure the robustness of the system. A comparison study was investigated to verify the performances of the DPC-based-BSC to that of the vector control (VC) based Proportional-Integral (PI). All were simulated using the Matlab/Simulink® environment.

Keywords DFIG · Wind turbine · Voltage dips · Grid faults · DPC approach · Backstepping technique · LVRT

S. Mensou (✉) · A. Essadki
Research Center of Engineering and Health Sciences and Technologies (STIS) of ENSAM, Mohammed V University, Rabat, Morocco
e-mail: sara.mensou11@gmail.com

Z. Mekrini
Laboratory of Informatics, Systems & Telecommunications, FST, Abdelmalek-Essaadi University, Tangier, Morocco
e-mail: zmekrini@uae.ac.ma

T. Nasser
Communication Networks Department of ENSIAS, Mohammed V University, Rabat, Morocco

B. B. Idrissi
Electromechanical Engineering Department of ENSAM, Moulay Ismail University, Meknes, Morocco

S. El Himer et al. (eds.), *Artificial Intelligence of Things for Smart Green Energy Management*, Studies in Systems, Decision and Control 446,
https://doi.org/10.1007/978-3-031-04851-7_4

1 Introduction

The wind system based on a DFIG has known a fast evolution in the last decades, for its inexpensive power electronic devices, robustness, and energy efficiency, which leads to an increasing insertion of this type of generator in wind systems connected to the electrical networks [1–3]. But during the grid faults, a destabilization of the wind systems is observed and its system protections operate to disconnect all the aero-generators from the electrical network, which leads to a loss of production of several tens of MW, and a destabilization of the electrical grid. In this context, the new grid codes remain the aero-generator connected to the network during the period of the fault to continue the production of the electricity [4].

As the stator of the generator is connected directly to the electrical grid, during a deep voltage sag, the stator currents increase, due to the strong magnetic coupling between the rotor and the stator windings, high peak currents appear on the rotor circuit and the electronic power converters, which leads to torque oscillations. Therefore, if the protection systems don't react quickly, the high peak currents can destroy the power converters and the DFIG [4–6].

Indeed, to protect wind systems and keep the aero-generator connected to the grid during and after the fault period, many control strategies have been investigated in the literature [7–19]; these solutions propose using protection devices or developing control methods for wind systems. The ability of the WTs to remain connected to the electrical network during the voltage dip is called Low Voltage Ride-Through Capability (LVRT).

The Crowbar protection circuit is one of the most widely used solutions to protect WTs against deep voltage sags, the principle of this technology is to connected a bank of resistors directly to the rotor of the generator when a voltage drop is detected, at which time the DFIG operates as a squirrel-cage machine [9–11], consequently the generator consume more reactive power and the stability of the network is lost. This solution limits the overcurrents limit in the rotor circuit, protects the generator and the power converters, but the controllability of the generator becomes impossible [12], then the control of the reactive power will be lost during the activation of the Crowbar circuit. Another solution that has been developed in [13–15] to compensate the reactive power; is the use of a Static Synchronous Compensator (STATCOM) in the wind system, this solution gives good results, but it is very expensive, therefore it increases the cost of the installing system.

As an alternative, a lot of control strategies have been proposed. In [16] a flux stator control approach based on the ADRC method associated with the Crowbar protection circuit has been improved, the flux control is applied to keep the magnetization of the generator under the grid fault, this solution works well, but when the Crowbar circuit was activated the controllability of the generator is lost. In articles [8–17] the DPC was applied to control the powers of the stator in order to protect the wind system during a Low Voltage Ride-Through (LVRT), the proposed control approach was validated by experiments results. The stator current control loop based on the Proportional-Integral (PI) corrector was presented in article [18], this control method was validated by laboratory experiments. Article [19] proposes a DPC-based on Sliding Mode Control under voltage sag; satisfactory results were obtained.

In this chapter we propose a new DPC method based on the nonlinear BSC to control the aero-generator under Low Voltage Ride-Through to:

- Contribute to the return of the voltage at its rated value.
- Perform the aero-generator operation.
- Suppress the high peak currents on the rotor and stator winding.
- Guaranty the electricity production during and after the grid fault.
- Reduce the total harmonic distortion (THD %) in the injected currents into the electrical network.

The applied regulator based on the BSC was designed using the mathematical model of the generator, and the flux stator will not be considered aligned to (d) axis and constant, then, its dynamics weren't be neglected in the synthesis of the regulators model. The reactive power will be controlled to contribute to the return of the voltage to its rated value, and the active power will be controlled to suppress the peaks and reduce the high amplitudes of the currents. To verify our improved method, the wind system model of rated power of 1500 W and DPC-based- BSC were compared whit PI controller under LVRT and simulated using Simulink® software.

This chapter is presented as follows: in Sect. 2 the definition and the Characterization of voltage dips are given. Section 3, the wind turbine and DFIG mathematical model were developed. The DPC based-BSC is presented in Sect. 4. In Sect. 5 the simulations results of the system are analyzed and discussed. Finally, in Sect. 6 some conclusions are given.

2 Voltage Dips

2.1 Definition and Characterization of a Voltage Dips

According to the European standard NF EN 50160 created by CENELEC (European Committee for Electrotechnical Standardization), which is currently applied in Morocco, a voltage dip is a sudden and temporary drop of the electrical network voltage, which can affect one or more phases in a point of the electrical network. The EN 50160 defines voltage dips for a drop of 10–90% in the RMS voltage for a period of 10 ms to one minute [5].

A voltage dip is characterized by its duration Δt and its amplitude ΔU. the ΔU is defined by the minimum value of the nominal voltage under the limit of the contractual interval (90% Un) in a cycle, thus the duration Δt is determined by the period between the passage of the voltage below the contractual limit (90% Un) and the return above this limit [20]. Figure 1 illustrates the characterization of a voltage dip.

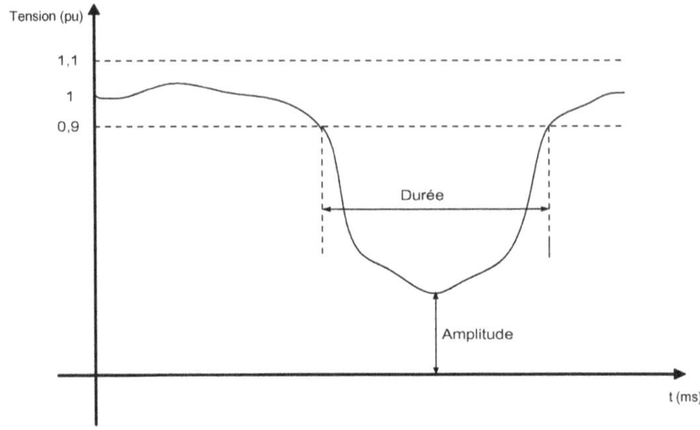

Fig. 1 Characterization of a voltage dips [18]

There are two types of voltage dips; asymmetrical voltage sag is when the voltage fault is not the same on the three phases, and symmetrical voltage dip is when the voltage fault is the same on all three phases. The number, duration, and amplitude of a voltage dip change from one grid connection point to another, and depend on the type of electrical network: underground or aerial, transmission or distribution.

2.2 *Origin of a Voltage Dips*

The amplitude of the voltage can be subject to abnormal variations and can even collapse to a level close to zero. Indeed, these types of frequent disturbances disrupt the operation of wind turbines and can cause unexpected shutdowns of these devices, which leads to a loss of production of several tens of MW and the deterioration of the equipment [3]. According to [20], we can distinguish several types of phenomena having consequences on voltage variations, they can occur at the producer:

- Accidental short-circuits.
- Meteorological phenomena such as lightning affect the electrical network or the connected installations.
- Disconnection of a production unit.
- Voltage drops due to the circulation of a large reactive power in the grid.

As they may occur to the consumer, namely:

- The connection of high loads which can cause voltage variations.
- Significant currents due to the starting of high power motors, electric arcs, or the saturation of transformers.

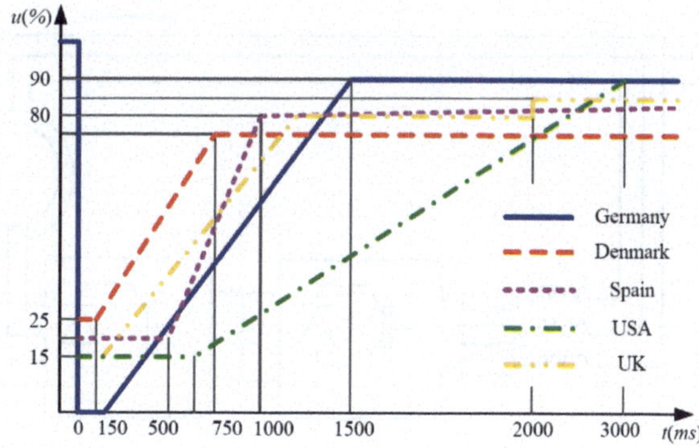

Fig. 2 Zone limits supported by the WT under voltage drops in several countries

2.3 Limit Zone of Voltage Dips Supported by WTs

Currently, new grid codes require wind power systems to remain connected to the electrical grid under voltage variations. Figure 2 shows the zone limits supported by a WT under voltage sags in some countries: the United Kingdom, the United States, Spain, Denmark, and Germany. The aero-generators must remain connected for a voltage dip above the limit zone defined by the grid code [21, 22].

When the RMS voltage goes below the critical limits zone presented, the protection systems disconnect the WTs from the grid, but when the voltage is above this limit, the WT must remain connected and could also contribute to the returns of the voltage to its nominal value.

3 Modeling of the Wind Energy System

The structure investigated in this chapter is based on the wind system used a DFIG with Gearbox; the stator of the DFIG is connected to the network, but the rotor is connected to the grid via two powers converters to operate on variable speed. A DC Bus capacitor is used between the converters to reduce the voltage ripples. Figure 3 presents the structure of the system studied.

3.1 Modeling of the Wind Turbine

The mathematical model of the WT is giving by these expressions [23, 24]:

$$P_v = \frac{\rho S v^3}{2} \tag{1}$$

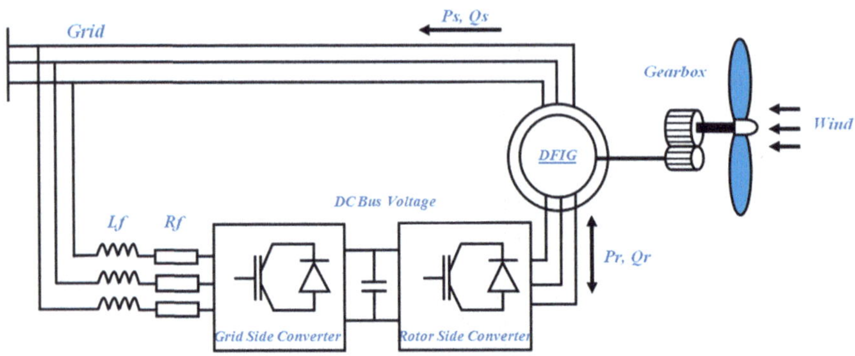

Fig. 3 Structure of the aero-generator

$$P_{aer} = \frac{\rho S v^3}{2} C_p(\lambda, \beta) \tag{2}$$

$$T_{aer} = \frac{P_{aer}}{\Omega_t} = \frac{\rho S v^3}{2\Omega_t} C_p(\lambda, \beta) \tag{3}$$

The Tip Speed Ratio (TSR) λ:

$$\lambda = \frac{\Omega_t R}{v} \tag{4}$$

where C_p, v, S, ρ, and β are the power coefficient, the wind speed, the circular surface swept by the turbine, the air density, and the pitch angle of the blade, respectively.

3.2 Modeling of the DFIG

The DFIG is modeled in the axis of the Park frame (d-q). The equations voltages of the machine can find their expression as [25, 26]:

$$\begin{cases} V_{s,dq} = R_s I_{s,dq} + \dfrac{d\varphi_{s,dq}}{dt} \pm \varphi_{s,qd}\omega_s \\ V_{r,dq} = R_r I_{r,dq} + \dfrac{d\varphi_{r,dq}}{dt} \pm \varphi_{r,qd}\omega_r \end{cases} \tag{5}$$

The expressions of the flux as a function of currents are given by:

$$\begin{cases} \varphi_{s,dq} = L_s I_{s,dq} + L_m I_{r,dq} \\ \varphi_{r,dq} = L_r I_{r,dq} + L_m I_{s,dq} \end{cases} \tag{6}$$

Considering the flux expressions in (6), we can find the expressions for the currents:

$$\begin{cases} I_{s,dq} = \frac{1}{\sigma L_s}\varphi_{s,dq} - \frac{k_r}{\sigma L_s}\varphi_{r,dq} \\ I_{r,dq} = \frac{1}{\sigma L_r}\varphi_{r,dq} - \frac{k_s}{\sigma L_r}\varphi_{s,dq} \end{cases} \tag{7}$$

$\sigma = 1 - \frac{L_m^2}{L_s L_r}$, $k_s = \frac{L_m}{L_s}$, $k_r = \frac{L_m}{L_r}$.

The apparent stator power of the DFIG in complex notation is provided by the following equation:

$$\overline{S_s} = P_s + jQ_s \tag{8}$$

The powers of the stator:

$$\begin{cases} P_s = V_{sd} I_{sd} + V_{sq} I_{sq} \\ Q_s = V_{sq} I_{sd} - V_{sd} I_{sq} \end{cases} \tag{9}$$

The electromagnetic torque of the machine considered [27]:

$$T_{em} = p\frac{L_m}{L_s}\left(\varphi_{sq} I_{rd} - \varphi_{sd} I_{rq}\right) \tag{10}$$

where p is the number of the pole pairs of the DFIG.

R_r, R_s: The rotor and stator resistance.

L_r, L_s: The cyclic rotor and stator inductance.

L_m: The mutual inductance.

4 DPC Control

The principle of the DPC (Direct Power Control) method is ensuring a direct and separate control of the powers generator. In each axis, one regulator is used to ensure separate and direct active and reactive powers control and eliminate the controlled variable errors. At the output of each regulator, a control voltage is calculated as a function of the stator flux, this voltage will be generated thereafter to control the power converters. The applied regulator is based on the Backstepping approach, with its use, the studied system can achieve high performance, and ensure its stability if our controller satisfies the conditions of Lyapunov stability [27, 28].

DPC-based-BSC is proposed to guarantee robust power regulation during voltage dips in the network. Our controller will be elaborated using the mathematical model of the double-fed induction generator. Substituting Eqs. (7) into (9), the active and reactive powers expressions become:

$$\begin{cases} P_s = \dfrac{k_r V_s}{\sigma L_s}\left(\dfrac{1}{k_r}\varphi_{sq} - \varphi_{rq}\right) \\ Q_s = \dfrac{k_r V_s}{\sigma L_s}\left(\dfrac{1}{k_r}\varphi_{sd} - \varphi_{rd}\right) \end{cases} \tag{11}$$

Using Eqs. (5), (7), and (11) we can find the expressions of the rotor voltages as a function of the fluxes:

$$\begin{cases} V_{rd} = -\dfrac{1}{\mu}\dfrac{dQ_s}{dt} - \alpha Q_s + \dfrac{1}{\mu}\omega_r P_s + \dfrac{1}{k_r}\dfrac{d\varphi_{sd}}{dt} + \delta\varphi_{sd} - \dfrac{1}{k_r}\omega_r\varphi_{sq} \\ V_{rq} = -\dfrac{1}{\mu}\dfrac{dP_s}{dt} - \alpha P_s - \dfrac{1}{\mu}\omega_r Q_s + \dfrac{1}{k_r}\dfrac{d\varphi_{sq}}{dt} + \delta\varphi_{sq} + \dfrac{1}{k_r}\omega_r\varphi_{sd} \end{cases} \tag{12}$$

where $\mu = \frac{k_r V_s}{\sigma L_s}$; $\alpha = \frac{R_r}{\sigma L_r \mu}$; $\delta = \frac{R_r(1-k_r k_s)}{\sigma L_r k_r}$.

The expressions for the rotor voltages in Eq. (12) contain six terms, the first three terms show the normal operation of the aero-generator and the last three terms present the flux variation under voltage drop. Equation (12) can be written in another form:

$$\begin{cases} \dfrac{dQ_s}{dt} = -\mu\alpha Q_s + \omega_r P_s + \dfrac{\mu}{k_r}\dfrac{d\varphi_{sd}}{dt} + \mu\delta\varphi_{sd} - \dfrac{1}{k_r}\mu\omega_r\varphi_{sq} - \mu V_{rd} \\ \dfrac{dP_s}{dt} = -\mu\alpha P_s - \omega_r Q_s + \dfrac{\mu}{k_r}\dfrac{d\varphi_{sq}}{dt} + \mu\delta\varphi_{sq} + \dfrac{1}{k_r}\mu\omega_r\varphi_{sd} - \mu V_{rq} \end{cases} \tag{13}$$

The errors of the stator powers:

$$\begin{cases} \varepsilon_Q = Q_s^* - Q_s \\ \varepsilon_P = P_s^* - P_s \end{cases} \tag{14}$$

The derivative of the errors:

$$\begin{cases} \dot{\varepsilon}_Q = \dot{Q}_s^* - \dot{Q}_s \\ \dot{\varepsilon}_P = \dot{P}_s^* - \dot{P}_s \end{cases} \tag{15}$$

Using Eq. (13), the derivative of the errors:

$$\begin{cases} \dot{\varepsilon}_Q = \dot{Q}_s * + \mu\alpha Q_s - \omega_r P_s - \dfrac{\mu}{k_r}\dot{\varphi}_{sd} - \mu\delta\varphi_{sd} + \dfrac{1}{k_r}\mu\omega_r\varphi_{sq} + \mu V_{rd} \\ \dot{\varepsilon}_P = \dot{P}_s * + \mu\alpha P_s + \omega_r Q_s - \dfrac{\mu}{k_r}\dot{\varphi}_{sq} - \mu\delta\varphi_{sq} - \dfrac{1}{k_r}\mu\omega_r\varphi_{sd} + \mu V_{rq} \end{cases} \tag{16}$$

We use the Lyapunov function to guarantee stability, it is expressed by:

$$v = \frac{1}{2}\varepsilon_Q^2 + \frac{1}{2}\varepsilon_P^2 \tag{17}$$

Substituting Eq. (16) into (17), the derivative of the Lyapunov function becomes:

$$\begin{aligned}
\dot{v} &= -k_Q\varepsilon_Q^2 - k_P\varepsilon_P^2 \\
\dot{v} &= \varepsilon_Q(\dot{Q}_s^* + \mu\alpha Q_s - \omega_r P_s - \frac{\mu}{k_r}\dot{\varphi}_{sd} - \mu\delta\varphi_{sd} + \frac{1}{k_r}\mu\omega_r\varphi_{sq} + \mu V_{rd}) \\
&\quad + \varepsilon_P(\dot{P}_s^* + \mu\alpha P_s + \omega_r Q_s - \frac{\mu}{k_r}\dot{\varphi}_{sq} - \mu\delta\varphi_{sq} - \frac{1}{k_r}\mu\omega_r\varphi_{sd} + \mu V_{rq})
\end{aligned} \tag{18}$$

To guarantee stability and obtain a good regulation of the stator powers, the gains k_Q and k_p must be positive and the derivative of v negative. According to (18), we can find the expressions of the control voltages:

$$\begin{cases}
V_{rd}^* = -\left(\frac{1}{\mu}\left(\dot{Q}_s^* + k_Q\varepsilon_Q - \omega_r P_s\right) + \alpha Q_s\right) - \frac{1}{k_r}(\omega_r\varphi_{sq} + \dot{\varphi}_{sd}) + \delta\varphi_{sd} \\
V_{rq}^* = -\left(\frac{1}{\mu}\left(\dot{P}_s^* + k_P\varepsilon_P + \omega_r Q_s\right) + \alpha P_s\right) + \frac{1}{k_r}(\omega_r\varphi_{sd} - \dot{\varphi}_{sq}) + \delta\varphi_{sq}
\end{cases} \tag{19}$$

We know that the expression of the maximum stator apparent power Ss is given by [9]:

$$S_{s-\max} = I_{sn}V_{sn} \tag{20}$$

where I_{sn} and V_{sn} are the rated stator current and voltage of the DFIG. By respecting the grid code, when the system detects a voltage dip for a duration Δt less than 1 s and with a depth d between 30 and 80%, the power references become:

$$\begin{cases}
Q_s^* = -I_{sn}V_{sq}\left(1 - \frac{V_{sq}}{V_{sn}}\right) \\
P_s^* = 0
\end{cases} \tag{21}$$

5 Simulation Results and Discussions

The simulation of the wind power system under a voltage drop was carried out using the Matlab/Simulink environment; the parameters of the system studied are given in Table 1. We assumed that the wind speed is constant with an average value of v = 11 (m/s) during the voltage drop, because the duration of voltage drops is smaller than the period of the fluctuation in the wind speed [5–16].

Table 1 Wind system parameters

Symbol	Quantity	Value
P_n	Nominal power	1.5 kW
V_s	Stator voltage	220/380 V
R_r	Rotor resistance	1.65 Ω
R_s	Stator resistance	1.18 Ω
L_r	Rotor inductance	0.17 H
L_s	Stator inductance	0.20 H
R	Turbine radius	1 m
G	Multiplier	4
J	Total inertia	0.04 kg m^2
C_f	Coefficient of frictions	0.007
ρ	Air density	1.22 kg/m^3

The wind turbine operates in MPPT mode and without disturbance with Qs = 0 (VAR) and Ps = −1350 (W), the reactive power is regulated at zero in to have a unit power factor on the stator side, the control is activated to allow detection of the voltage sag, a sudden voltage drop appears on the stator power supply terminals at t = 3 (s), consequently, the BSC calculates and generates the values of the rotor control voltages.

We apply to the terminals of the stator power supply a balanced voltage dips with a depth of 60% for a duration of 500 (ms) which appears at t = 3 (s) and disappears at t = 3.5 (s) (see Fig. 4). Figure 5 illustrates the evolution of the currents stator $I_{s\text{-}abc}$ during and after the fault grid, we can see that it increase all the period of the dip on the stator voltages.

The system control calculates the values of the reference powers, the reactive power becomes Q_s = -800 (VAR) during the fault (see Fig. 6), the power Qs is transferred into the electrical network to contribute to the return of the voltage to its rated value, the active power shown in Fig. 7 becomes P_s = 0 (W) to suppress the overcurrents limits. After the fault, the wind generator continues its normal operation

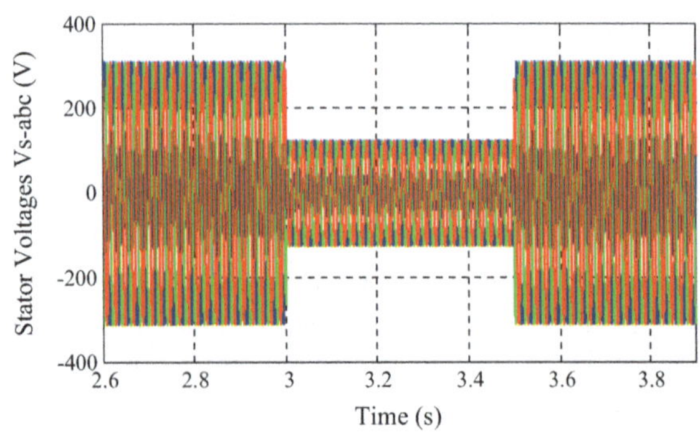

Fig. 4 Stator voltages under voltage dips (ΔVs = 186 V)

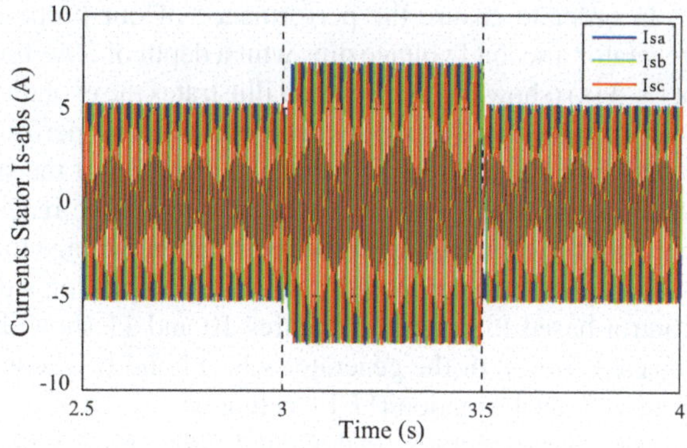

Fig. 5 Stator currents $I_{s\text{-}abc}$ (A)

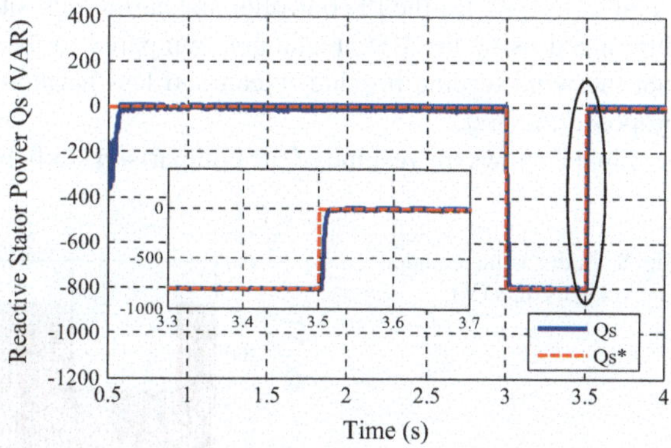

Fig. 6 Reactive stator power Q_s (VAR)

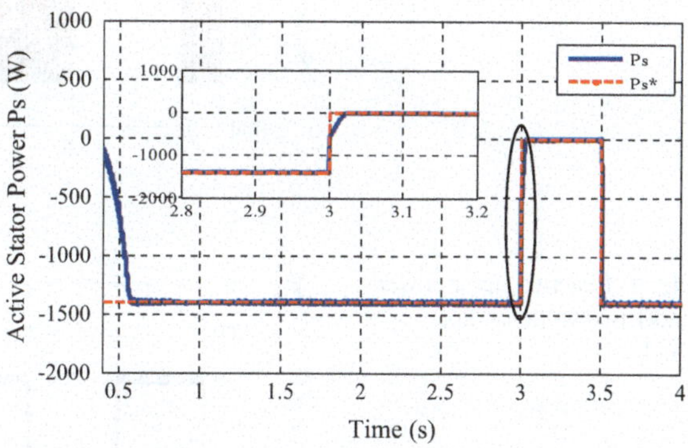

Fig. 7 Active stator power P_s (W)

and reproduces electrical power. As we observe in Figs. 6 and 7, the monitoring of the command setpoint is perfectly performed during and after the fault, the powers are well decoupled, the BSC controller reaches the references of the stator powers; with a fast response time, without power oscillations on appearance and disappearance of the voltage dip.

In order to ensure the performance of our proposed method, we apply to the generator a second voltage dips whit a depth of 30% for a duration equal to 400 (ms) at t = 5 (s) (show Fig. 8). Figure 9 illustrates the evolution of the reactive power under two voltages drop, as we see the good tracking performances is well done, and the reactive power changes its value as a function of the stator voltages (show Eq. 21), which confirm that the aero-generator inject the reactive power into the electrical network for contributing to the return of the voltage to it rated value.

To validate our proposed method, a comparison study was done whit the vector control-based PI corrector. Figures 10 and 11 show the harmonic spectrum of the injected current of the generator whit PI and BSC controllers, obtained by the Fast Fourier Transformation (FFT) technique.

We observe that the fundamental frequency is equal to 50 Hz for both methods, the Total Harmonic Distortion is equal to 4.12% for the DPC-BSC controller, and equal to 15.09% for the PI controller, the harmonics found in the current injected are minimized using the BSC technique compared to the PI controller, which implies that the wind turbine provides clean and less harmonic electrical energy with our proposed method.

Table 2 shows the resume of the comparison study between our proposed method

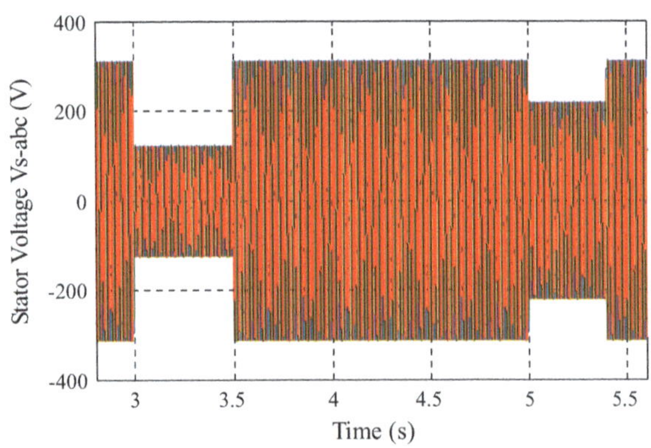

Fig. 8 Stator voltage under two voltages dips (V)

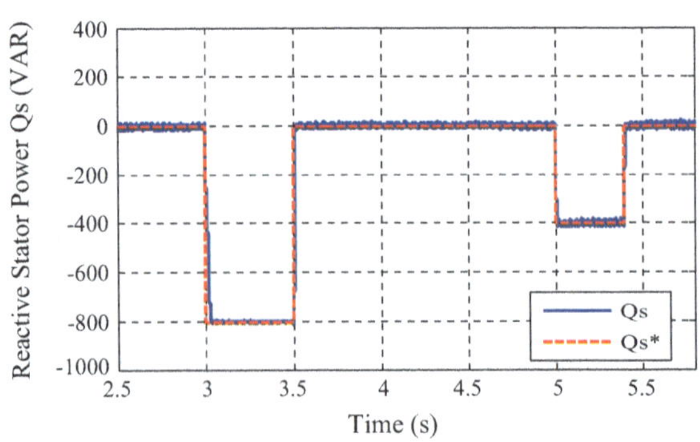

Fig. 9 Reactive stator power under two voltages dips (VAR)

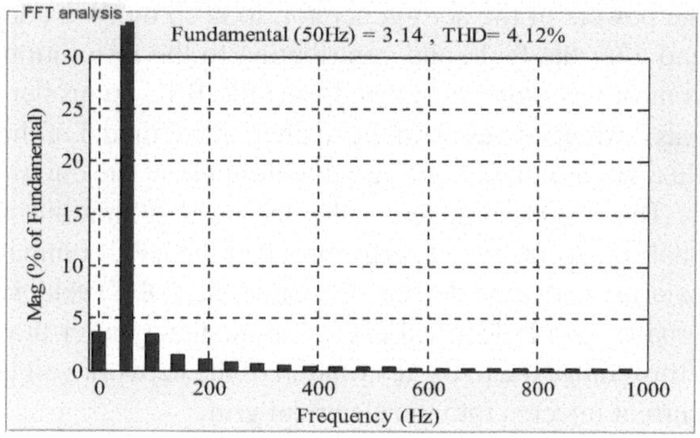

Fig. 10 Spectrum analysis of the THD for the current stator I_{sa} whit DPC-BSC during grid fault

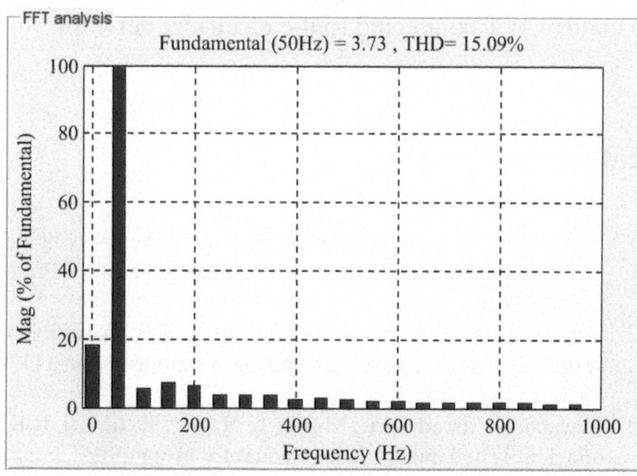

Fig. 11 Spectrum analysis of the THD for the current stator I_{sa} whit VC-PI during grid fault

Table 2 Comparison between our proposed approach and the VC-PI controller under voltage dips

Controller	Rise time (s)	Response time (s)	THD (%) of the Current I_{sa}	Robustness
DPC-BSC	**0.13**	**0.15**	**4.12**	**Robust**
VC-PI [28]	0.20	0.38	15.01	Not robust

and the vector control-based PI corrector in terms of rise time, response time, robustness, and THD (%) on the current of the DFIG. As we see the DPC-BSC approach is robust, has a faster rise and response time compared to VC-PI, and the harmonics detected on the current are minimized.

6 Conclusion

In this chapter, we have presented the modeling of the wind system based-DFIG, and then we have developed a control method using the DPC command to control the powers of the aero-generator, to keep the WT connected to the network during and after the fault, and contributing to the restoration of the network voltage. The control technique developed uses the BSC controller as a regulator; this regulator was developed based on the mathematical model of the generator, in which its stator flux was assumed to be variable throughout the duration of the fault.

The results obtained under the Matlab/Simulink software were presented and analyzed to verify the performance of the DPC command. It can be said that using an external hardware device, the command (DPC) achieved its objectives; continue the production of electrical energy immediately after the disappearance of the voltage drop, contribute to the restoration of the network voltage, and reduce the THD of the current injected into the electrical grid.

According to the simulation result, we can conclude that the DPC control method based on the Backstepping controller is a robust, efficient, and appropriate method to control WT connected to the electrical network.

References

1. Nadour, M., Essadki, A., Nasser, T.: Improving low-voltage ride-through capability of a multi-megawatt DFIG based wind turbine under grid faults. Protect. Control Mod. Power Syst. **5**(1), 1–13 (2020)
2. Mensou, S., Essadki, A., Nasser, T., Idrissi, B.B.: An efficient nonlinear Backstepping controller approach of a wind power generation system based on a DFIG. Int. J. Renew. Energy Res. **7**(4), 1520–1528 (2017)
3. Mensou, S., Essadki, A., Minka, I., Nasser, T., Idrissi, B.B.: Control and hardware simulation of a doubly-fed induction aerogenerator using dSPACE card. In: 2nd International Conference of Computer Science and Renewable Energies (ICCSRE), Agadir-Morocco, 22–24 (July 2019)
4. Jacomini, R.V., Sguarezi Filho, A.J.: Finite control set applied to the direct power control of a DFIG operating under voltage sags. IEEE Trans. Sustain. Energy **10**(2), 952–960 (2019)
5. Mensou, S., Essadki, A., Nasser, T., Idrissi, B.B.: A direct power control of a DFIG based-WECS during symmetrical voltage dips. Protect. Control Modern Power Syst. **5**(5) (2020)
6. Peng, L.: Reconfiguration du dispositif de commande d'une éolienne en cas de creux de tension. Ph.D. thesis No 122, Central School of Lille, Jun (2010)
7. Mekrini, Z., Bri, S.: Performance of an indirect field-oriented control for asynchronous machine. Int. J. Eng. Technol. **8**(2), 726–733 (2016)
8. Jacomini, R.V., Sguarezi Filho, A.J.: Direct power control strategy to enhance the dynamic behavior of DFIG during voltage sag. In: Proceeding of the 7th International Conference on Renewable Energy Research and Applications (2018), pp. 194–198
9. Franco, R., Capovilla, C.E., Jacomini, R.V., Altana, J.A.T., Sguarezi Filho, A.J.: A deadbeat direct power control applied to doubly-fed induction aerogenerator under normal and sag voltages conditions. In: Proceeding of the 40th Annual Conference of the IEEE Industrial Electronics Society (2014), pp. 1906–1911
10. Morren, J., de Haan, S.W.H.: Ride through of wind turbines with doubly-fed induction generator during a voltage dip. IEEE Trans. Energy Convers. **20**(2), 435–441 (2005)
11. Olea, E., Ruiz, J., Lopez, J., Gubia, E., Marroyo, L.: Ride through of wind turbines with doubly fed induction generator under symmetrical voltage dips. IEEE Trans. Industr. Electron. **56**(10), 4246–4254 (2009)
12. Xiao, S., Yang, G., Zhou, H., Geng, H.: A LVRT control strategy based on flux linkage tracking for DFIG-based WECS. IEEE Trans. Industr. Electron. **60**(7), 2820–2832 (2013)

13. Reddy, G.M., Manohar, T.G.: Fuzzy logic controller based STATCOM for grid connected wind turbine system. Int. J. Renew. Energy Res. **8**(2), 702–713 (2018)
14. Qiao, W., Venayagamoorthy, G.K., Harley, R.G.: Real-time implementation of a STATCOM on a wind farm equipped with doubly fed induction generators. IEEE Trans. Ind. Appl. **45**(1), 98–107 (2009)
15. Thirupathaiah, M., Prasad, P.V., Ganesh, V.: Enhancement of power quality in wind power distribution system by using hybrid PSO-firefly based DSTATCOM. Int. J. Renew. Energy Res. **8**(2), 1138–1154 (2018)
16. Chakib, R., Cherkaoui, M., Essadki, A.: Stator flux control by active disturbance rejection control for DFIG wind turbine during voltage dip. Int. J. Circuits Syst. Signal Process. **9**, 281–288 (2015)
17. Solís-Chaves, J.S., Barreto, M.S., Salles, M.B., Lira, V.M., Jacomini, R.V., Sguarezi Filho, A.J.: A direct power control for DFIG under a three phase symmetrical voltage sag condition. Control Eng. Pract. **65**, 48–58 (2017)
18. Francisco, K., Lima, A., Luna, A., Rodriguez, P., Watanabe, E.H., Blaabjerg, F.: Rotor voltage dynamics in the doubly fed induction generator during grid faults. IEEE Trans. Power Electron. **25**(1), 118–130 (2010)
19. Jeong, H.G., Ro, H.S., Lee, K.B.: A control scheme of the low voltage ride through in wind turbines using a direct power control based on a sliding mode control. In: Proceeding of the International Symposium on Industrial Electronics (2013), pp. 1–6
20. Chakib, R.: Commande avancée d'une éolienne à base de la MADA en vue de sa participation aux services système: réglage de fréquence, réglage de tension et tenue aux creux de tension. Ph.D. thesis, ENSET-Mohammed 5 university of Rabat (2017)
21. Teninge, A.: Participation aux services système de parcs éoliens mixtes: application en milieu insulaire. Ph.D. thesis, National Polytechnic Institute of Grenoble, French (2009)
22. Lima, F.K.A., Luna, A., Rodriguez, P., Watanabe, E.H., Blaabjerg, F.: Rotor voltage dynamic in the doubly fed induction generator during grid faults. IEEE Trans. Power Electron. **25**(1) (2010)
23. Mensou, S., Essadki, A., Minka, I., Nasser, T., Idrissi, B.B. : Control and hardware simulation of a doubly fed induction aero-generator using dSPACE card. In: Proceeding of the 2nd International Conference of Computer Science and Renewable Energies (2019)
24. Mensou, S., Essadki, A., Minka, I., Nasser, T., Idrissi, B.B., Bentarla, L.: Performance of a vector control for DFIG driven by wind turbine: real time simulation using DS1104 controller board. Int. J. Power Electron. Drive Syst. **10**(2), 1003–1013 (2019)
25. Kharchouf, I., Essadki, A., Nasser, T.: Wind system based on a doubly fed induction generator: contribution to the study of electrical energy quality and continuity of service in the voltage dips event. Int. J. Renew. Energy Res. **7**(4), 1892–1900 (2017)
26. Mekrini, Z., Seddik, B.: A modular approach and simulation of an asynchronous machine. Int. J. Electr. Comput. Eng. **6**(4), 2088–8708 (2016)
27. Mensou, S., Essadki, A., Nasser, T., Idrissi, B.B.: A robust speed control of a doubly-fed induction generator using in WECS by the nonlinear Backstepping controller. In: 3rd International Conference on Electrical and Information Technologies (ICEIT), Rabat-Morocco, 15–18 November 2017
28. Mensou, S.: Commande et amélioration des performances d'un aérogénérateur fonctionnant a vitesse variable a base de la génératrice asynchrone doublement alimentée. Ph.D. thesis, ENSAM- Mohammed V university of Rabat (2021)

Optimization of Two Hybrid Micro-Concentrator Photovoltaic Systems for Car-Roof Application

Sarah El Himer, Mariyam Ouaissa, and Mariya Ouaissa

Abstract This paper covers a general, parametric and performance analysis of tow optical elements to be used in micro-concentrator photovoltaic (MCPV) systems for car roof application using the same primary optical element which is the Fresnel lens (FL). It concerns a non-imaging optic which is: Pyramid and an imaging optic: square dome. The performance of all concentrator systems is compared in terms of actual concentrations achieved, angles of acceptance and output light power distribution. The results are compared with the same types of SOs but are designed for car roof applications. For the first simulation results, we found that the Fresnel lens with 28 prisms and a focal length of 14 mm, is the optimal primary optic, and the pyramid is the most suitable element as a secondary optical element placed at focal. By adding the silicon solar cell to multijunction solar cell, the efficiency of the optical system is increased by 20%. Therefore the power generated is 41.8 kw and according to Table 2 therefore this car can drive with a speed of 120 km/h.

Keywords Primary optic · Secondary optic · Micro CPV · Fresnel lens · Flux uniformity acceptance angle · Optical efficiency · Pyramid · Dome

1 Introduction

The electric vehicle is the key element of green and ecological transport. It partially solves problems related to energy and the environment [1]. An electric vehicle is a vehicle powered by an electric motor. This motor is powered by a rechargeable

S. El Himer (✉)
Sidi Mohammed Ben Abdellah, Fez, Morocco
e-mail: sarah.elhimer@usmba.ac.ma

M. Ouaissa · M. Ouaissa
Moulay Ismail University, Meknes, Morocco
e-mail: mariyam.ouaissa@edu.umi.ac.ma

M. Ouaissa
e-mail: mariya.ouaissa@edu.umi.ac.ma

S. El Himer et al. (eds.), *Artificial Intelligence of Things for Smart Green Energy Management*, Studies in Systems, Decision and Control 446,
https://doi.org/10.1007/978-3-031-04851-7_5

battery or by a fuel cell, or even by a thermal generator engine. Fully electric vehicles (Battery Electric Vehicle or BEV). The energy sources for this type of vehicle are batteries that can be recharged by the electricity grid or by other renewable energies (solar, wind turbines, etc.), among the solar technologies suitable for this application are concentration photovoltaics (CPV) and Micro CPV.

Concentrated Photovoltaic (CPV) is one of the most promising solar-energy-based methods for generating sustainable electricity. It is a high-efficiency replacement for the conventional flat plate module [2]. This method provides for an important reduction in the amount of space required in the cells, the recent halt of activity of several CPV players has posed a challenge to producing a given amount of electrical power while lowering its cost through high efficiency and moderate expenses. This commercial failure contrasts with the technological success that has been attained in the field, with modules nearing 40% efficiency [3].

The goal of employing micro-CPV is to lower cell size to the sub-mm range in order to increase a variety of benefits while also presenting certain production problems due to the small cell size. The micro-CPV shrinks CPV modules since their thickness, weight, and volume are related to the cell size; hence, the module's width is reduced.

One or two optical components can be used to make concentration optics. Several writers have utilized various Fresnel lenses [4–15], parabolic mirrors [16], and compound parabolic concentrators [17, 18]. The results demonstrate that using a single stage for CPV systems can't meet the general performance criteria; either the system is extremely efficient but the flux distribution on the cell isn't homogenous, or the converse is true. Then multistage concentrators, which are concentration systems with two optical components, the first of which is known as the Primary Optic (PO) and the second as the Secondary Optic (SO), can be more efficient. According to the literature, For a 1000× desired concentration ratio, we discovered various systems based on the Fresnel lens, with secondary systems such as the Pyramid, the Crossed Compound Parabolic Concentrator, Dome and Hyperbole, which were explored by Rodrguez et al. [19, 20]. The four concentrators have an acceptance angle of less than 1° and an optical efficiency of more than 80%, according to simulation findings. With the pyramid, the best flux uniformity may be achieved. Experiments confirmed these findings, with the optical efficiency falling to 73% and the acceptance angle falling to 0.8°. El Himer et al. [21] evaluated and assessed four CPV systems based on four secondary optical components (CPC, Cone, CCPC, and pyramid) connected with a circular flat Fresnel lens for a 1000× as concentration ratio using just simulation. They also demonstrated that, regardless of the material employed, the pyramid as SO provides the most uniform lighting on the solar cell, the widest acceptance angle (1.4°), and the maximum optical efficiency (83%). The performance of parabolic mirrors as a main or secondary optic has been studied by several writers [22, 23]. The results demonstrate that using a parabolic mirror as a major optical element produces intriguing optical results, especially when combined with a flat mirror, as it achieves the highest optical efficiency of 85 percent. When two optical components of concentration are used, the CPV system is more efficient because it allows for high optical efficiency, high concentration, and a broad acceptance angle.

Multiple systems have been investigated in the MCPV system for several years [24, 25], and recent advancements have led to improved micro-concentrator ideas and modules [26, 27]. Ritou [28] investigated a system consisting of a 19 × 19 mm^2 conventional Plano convex lenses as the main optic (PO) and a secondary optic (SO), a dome lens with 0.6 mm2 cells as the secondary optic (SO) and a concentration factor of 1000×. Its findings revealed that the acceptance angle was around 0.7 degrees [29]. With a Plano convex lens POE of 25 mm and a cell aperture of 0.25 mm, Jared MCPV design in [29] reaches 100 with an acceptance angle of 0.57°. The diameter of each square lens in the font lens array in the Partric research is 2.0 mm, and each multi-junction cell is 0.0632 mm × 0.0632 mm, giving the system a geometric concentration of 0.0632 mm of 1000×. These results show that the system efficiency is about 92% while the acceptance angle is 0.8° [30]. Salima et al. [31] offered a comparison of two MCPV systems: the first uses a Fresnel lens as the main optic (PO) and a dome lens as the secondary optic (SO). The second is made up of two lenses. The primary is Plano-convex, while the secondary is formed like a dome. Polymethyl methacrylate (PMMA) and polycarbonate are two potential materials (PC). The results reveal that the system consisting of a Polymethyl methacrylate (PMMA) Plano convex lens and a dome has the best optical efficiency in terms of flux non-uniformity; nevertheless, if the two systems are built of Polycarbonate, they allow for a broad acceptance angle (PC).

The purpose of using the micro-CPV is to reduce the cell size to the sub-mm range in order to achieve a wide range of benefits, at the expense of introducing some manufacturing challenges related to the small size of the cells. The micro-CPV reduces the size of the CPV modules because their thickness, weight and volume are proportional to the size of the cell; therefore the width of the module is reduced.

2 Physical Variables Definition

Approximately 70% of passenger cars travel fewer than 30 km each day, according to literature, infrastructure, and transportation and tourism. A commercial plug-in hybrid car's electrical power consumption rate is now 8.8 km/kWh. When the weight of these cars is decreased to 600 kg, it is expected to improve to 17.0 km/kWh. Under standard irradiation circumstances in Morocco, a photovoltaic (PV) module with a rated power of 800 W installed in such cars generates an average generated electric power of 1.8 kWh/day, which is comparable to a driving range of 31 km/day. To create this amount of energy using a 2.5 m^2 area-constrained car roof (Fig. 1).

In the case, the dimensioning is done by considering that the vehicle is traveling on a flat road at a constant speed (zero acceleration). We consider several speed values ranging from 10 to 120 $km\ h^{-1}$.

Given the characteristics of the EV (BMW i3 (2019)), we determine the power and energy required for the vehicle. Note that the calculated power relates to any vehicle traveling at constant speed, whether it is an electric vehicle, gas vehicle, petroleum vehicle, or other type of vehicle. For example, a vehicle traveling at 50 km/h with a

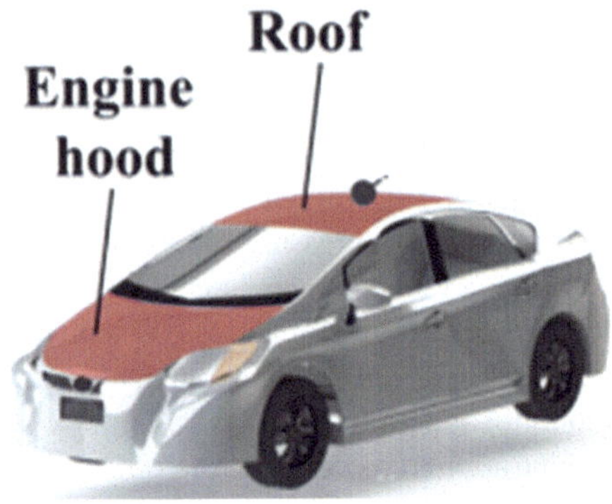

Fig. 1 Car roof for application

Table 1 Specifications of the car considered in the study

Maximum mass	1000 kg
Average speed	45 km/h
Maximum speed	120 km/h
Acceleration 0–45 km/h	less than 15 s
Autonomy	100 km
Coefficient of penetration in air	0.3
Coefficients of friction	0.01
Frontal area	2.5 m^2
Air density	1.225 kg/m^3
Acceleration of gravity	9.81 N/kg

range of 100 km needs 5.3 kWh of energy. On the other hand, for the same vehicle traveling at a speed of 120 km/h, the energy required is 17.2 kWh (Tables 1 and 2).

Table 2 Energy calculation result at constant speeds

Vehicle speed	10	30	50	90	110	120	Km.h-1
Speed	2.8	8.4	14	25.2	30.8	33.3	m.s-1
Rolling resistance	0.3	0.8	1.4	2.5	3	3.3	kW
Air resistance	0.01	0.3	1.3	7.4	13.4	17.3	kW
Total resistance force	101.7	130.5	188.1	389.8	533.8	518.6	N
Absorbed power	0.3	1.9	2.6	9.9	16.4	21	kW
Energy per 100 km	2.8	3.7	5.3	10.9	15	17.2	kWh

3 Simulation Results

3.1 Optimization of Fresnel Lens

The Fresnel lens is made of a series of narrow concentric grooves. In this part, we will consider the case of a two-dimensional Fresnel lens with grooves inward. Based on Fig. 2, we can draw the following equations:

$$\beta = r' + \theta_i \tag{1}$$

$$\theta_i = artg\frac{d}{2f} \tag{2}$$

$$\tan r' = \frac{d}{2(n\sqrt{\left(\frac{d}{2}\right)^2 + f^2 - 1)}} \tag{3}$$

The above equations allow us to design the 3D models of the Fresnel lens with inward grooves shown in Fig. 3 where we have taken as an example a Fresnel lens of 14 mm in diameter, a focal length of 14 mm and of 14 grooves.

a. **Nombre de prisme**

Chromatic aberration is caused by the angle at which sunlight refracts from the Micro CPV module's primary optical element (POE) depending on the wavelength [28]. Furthermore, due to current restrictions in the top cell and restricted semiconductor bandgap energy, infrared photons with wavelengths greater than 1.3 m are not

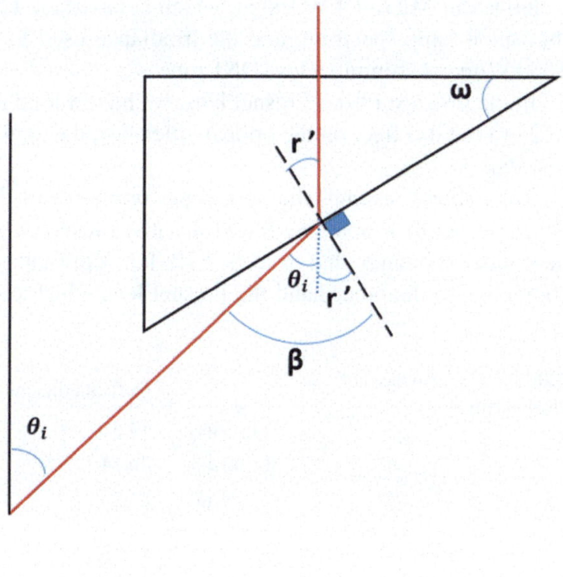

Fig. 2 Prism of the Fresnel lens oriented inward

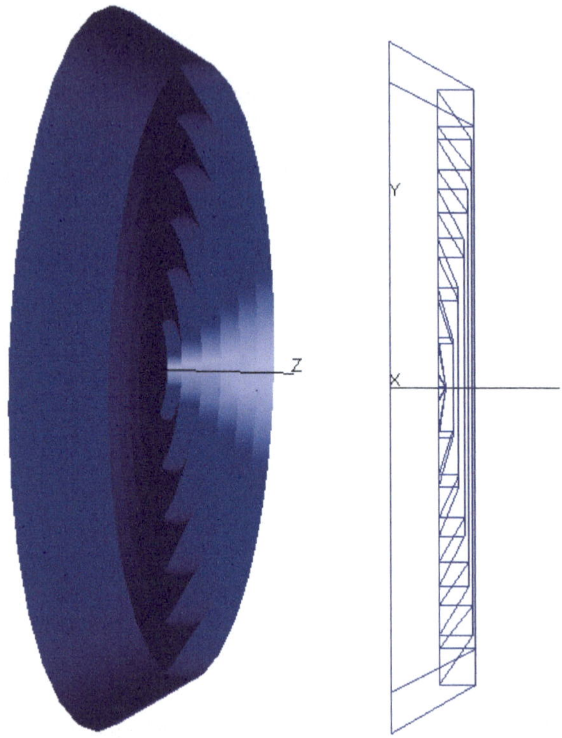

Fig. 3 Designed Fresnel lens

absorbed in either TJ solar cells or silicon sun cells. As a result, when optimizing the SOE of the micro CPV module, we studied the wavelength range of 0.3–1.3 m. We utilized the TracePro (Lambda Research Corporation) raytracing software to examine our Micro CPV design, which is based on a Monte Carlo approach using the xenon lamp spectrum, and the irradiance used to test the optical efficiency is 1000 W/m^2. Assuming a low-DNI zone.

In the first test of our Fresnel lens, we have varied the number of prisms from 7 to 28 to see its effect on the optical efficiency, the angle of acceptance and the flow distribution.

The Table 3 resumes the optical performances of the Fresnel lens proposed, as illustrated in this table, we found that the Fresnel lens with 28prisms could have an optical efficiency that exceeds 87% but it presents the lowest acceptance angle (0.6°) but on the other hand the Fresnel lens which contains 7 prisms presents the

Table 3 Performances of fresnel lens

	Optical efficiency (%)	Acceptance angle (°)
7 prisms	19.51	0.8
14 prisms	46.14	0.7
28 prisms	87.26	0.6

weakest optical efficiency but the widest angle of acceptance (0.8°). Regarding the distribution of the flux, we notice from Fig. 4 that whatever the number of prisms, the flux is concentrated with a high intensity in the center of the cell. From 1×10^{11} W/m^2 to 4×10^{10} W/m^2, the intensity drops. It should be noted that this diagram depicts the receiver's center intensity. When it comes to optical efficiency, the more prisms you have, the better. In fact, for a fixed diameter lens, decreasing the number of prisms increases the angle of the prism, and as a result, the size of the prism increases, resulting in losses from total internal reflection and geometrical losses (losses caused by the divergence of the optical beam) for rays that coincide with the ends of the prisms.

b. **Longueur focale**

We tested the optical performance of our Fresnel lens by changing the focal length from 7 to 28 mm in this section. The solar spectrum utilized at this time (Visible spectrum) is expanded from 300 to 1400 nm, and the prism number is set to 28 prisms.

Table 4 summarizes the optical performances of the proposed Fresnel lens. As shown in this table, the Fresnel lens with focal lengths of 14 and 28 mm has an optical efficiency of over 87 percent but has a low acceptance angle (0.6 and 0.35°), whereas the Fresnel lens with a focal length of 7 mm has the lowest optical efficiency but the widest acceptance angle (2.1°), These results are due to the fact that some light

Fig.4 Flux distribution

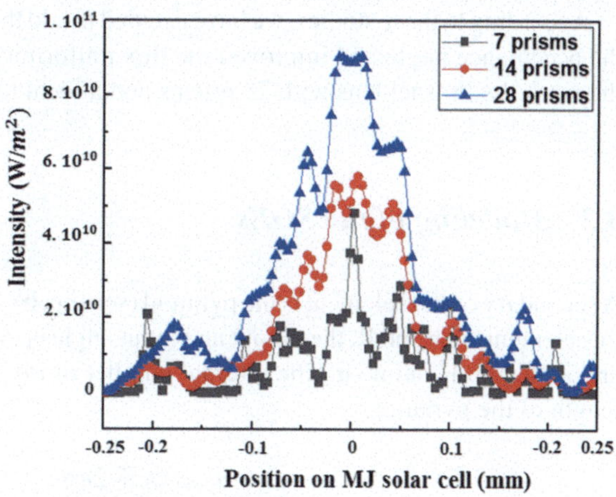

Table 4 Performances of Fresnel lens

	Optical efficiency (%)	Acceptance angle (°)
7 mm	19.37	2.1
14 mm	87.26	0.6
28 mm	87.8	0.35

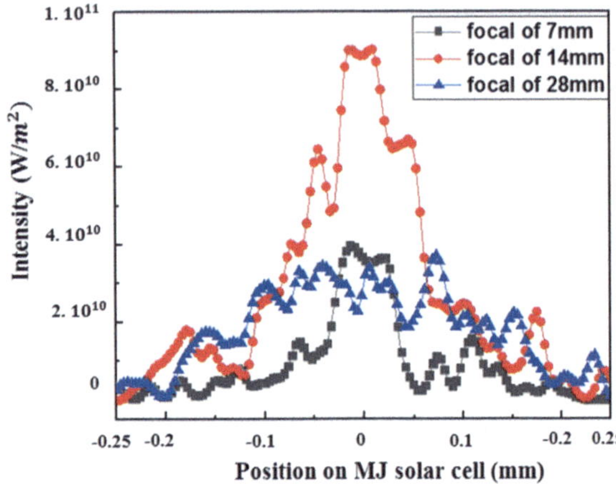

Fig. 5 Flux distribution

is reflected by TIR and does not reach the receiver at low focal lengths, but at large focal lengths, all rays reach the receiver.

The development of the intensity distribution on the receiver is seen in Fig. 5 for a variety of focal lengths. In the case of intensity, the intensity reaches a maximum for 14 mm as focal length (f-number from 1), and decreases with increasing focal length, but the perceived peak is wider, showing some homogeneity of the intensity repartition on the receiver.

According to these studies, we have decided to add the secondary optics to increase the acceptance angle and improved the flux uniformity. We consider for the rest of the work the Fresnel lens with 28 prisms and a focal length of 14 mm.

3.2 Modeling of the SOEs

A secondary optical element is the pyramid (see Fig. 6). The output and input radius of this element are a and R, the output and input angle θ_f and θ_i the angle of the pyramid, and its length H define it. The internal number of ray reflections, n, determines the length of the pyramid.

$$\theta_f = \theta_i + 2n\alpha \tag{4}$$

$$R = a + \sum_{n=1}^{n_{\max}} a_n \tag{5}$$

$$H = \frac{R - a}{\tan \alpha} \tag{6}$$

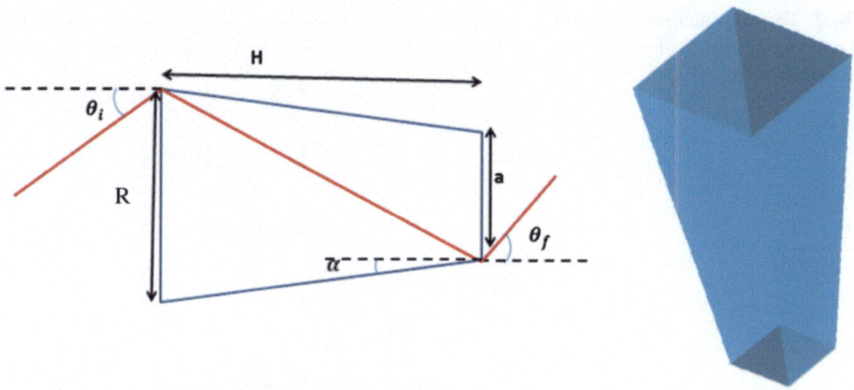

(a) *2D view of the Cone/Pyramid* **(b)** *Pyramid in 3D*

Fig. 6 Schematic view of cone/pyramid

The SO presenting is the dome with square exit as an imaging secondary optical element, The aim behind this SO (see Fig. 6) is to build its entrance surface in such a way that every ray.

In this section we present the dome with square exit as an imaging secondary optic, The objective behind this SO (see Fig. 5) is to construct its entry surface in such a manner that every ray coming from the main lens $A(x_1, y_1)$ hits the cell B's opposite extreme point (x_2, y_2). Furthermore, all rays that travel through point A and B, regardless of their angle of incidence, must follow Fermat's principle (Eq. 15) of optical path length conservation [31]. In these circumstances, Fermat's premise is as follows

$$l_1 + n_{SOE} l_{21} = cst \tag{7}$$

$$l_1 = \sqrt{(x - x_1)^2 + (y - y_1)^2} \tag{8}$$

$$l_2 = \sqrt{(x - x_2)^2 + (y - y_2)^2} \tag{9}$$

The beginning point for constructing the profile is the dome height h_{SO}. It is fixed by forcing the bidimensional profile's tangent to be horizontal at the origin, where F/# is the main lens's f-number (Fig. 7).

$$(x_0, y_0) = (0, h_{SOE}) = \frac{x_0}{\tan\left[arcsin\left(\frac{\mathrm{SIN}\left(arctg\left(\frac{1}{2F/\neq}\right)\right)}{n_{SOE}}\right)\right]} \tag{11}$$

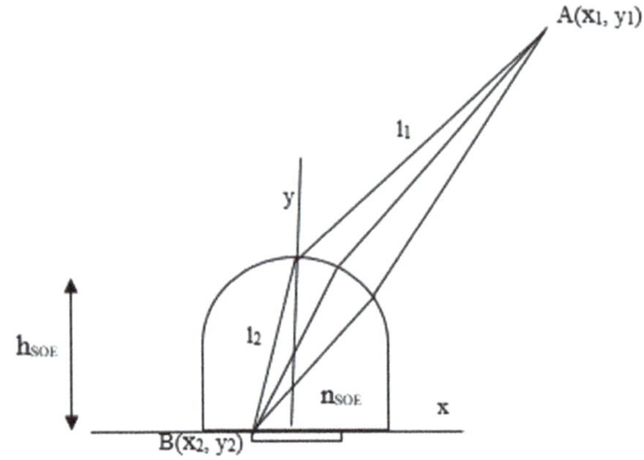

Fig. 7 Fermat principle applied to the design of secondary

3.3 Simulation Results

The study starts from optical modeling of the Fresnel lens and considers several sizes and geometric needs of SOs, We consider a Fresnel lens with 28 prisms and a focal length of 14 mm because we have obtained an optical efficiency of 87.26% and an acceptance angle of 0.6°.

The pyramid and dome were placed on the TJ solar cells for this test. We fixed the SO entry and kept the distance between the Fresnel lens and the TJ solar cell constant at this point. We computed the irradiation of the TJ solar cell using this SO. We also divided the results into an effective wavelength range (0.3–1.3 m) to assess the irradiance. Table 3 shows the optical performance of the pyramid and the dome. We notice that the micro-CPV system using a Fresnel lens associated to a dome presents high performances concerning the optical efficiency and the acceptance angle. Nevertheless, this SO presents a non-uniform flux distribution as shown in Fig. 8. In fact, the optical efficiency presented by pyramid is due to the losses of

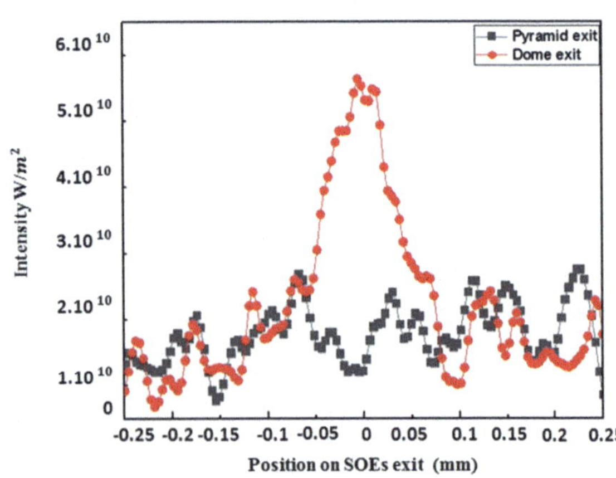

Fig. 8 Flux distribution on SOs exit

Table 5 Optical efficiency and acceptance angle of Micro CPV module with different SOs, as calculated using TracePro

Heights (mm)	Pyramid		Hyperbole	
	Optical efficiency (%)	Acceptance angle (°)	Optical efficiency (%)	Acceptance angle (°)
0.8 mm	79.3	0.9	86.76	1

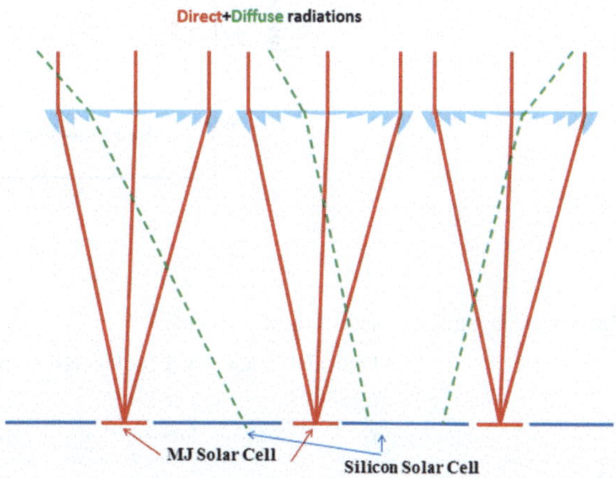

Fig. 9 Schematic view of hybrid CPV architecture

the rays concentrated by the Fresnel lens in fact, the surface of the spot of the rays converged by the Fresnel lens is greater from the entrance surface of the pyramid (Table 5).

We put a silicon sun cell of ($100 \times 100\ mm^2$) under the MJ solar cell to recover the rays that fall out of the dome and improve the performance of the mini CPV system, as shown in Fig. 9. To evaluate the performance of this hybrid micro-CPV system, we model it at various incidence angles (ranging from 0° to 20°). Figure 10 depicts the optical efficiency that can be achieved using silicon solar cells. We notice that the diffuse light which arrives with inclinations, can reach silicon cells depending on the in cadence angle as follows:

For $0° \leq \theta \leq 2°$, part of light reaches the dome exit and the other part falls on Silicon cells, and the optical efficiency reaches the 78%

$\theta = 0°$, all rays are collected at the dome output then the TJ solar cell and the optical efficiency reaches the 83%

Table 4 presents a comparison of the optical performances of the hybrid micro CPV to micro CPV system. We can see that hybrid micro CPV was able to increase the power by 20% (Table 6).

By checking if this power could run a car at any speed, and by exploiting the roof surface of the car, we need 250 silicon cells under 250 multi-junction cells. Therefore the power generated is 205kw and according to Table 2 therefore this car can drive with a speed of 120 km/h.

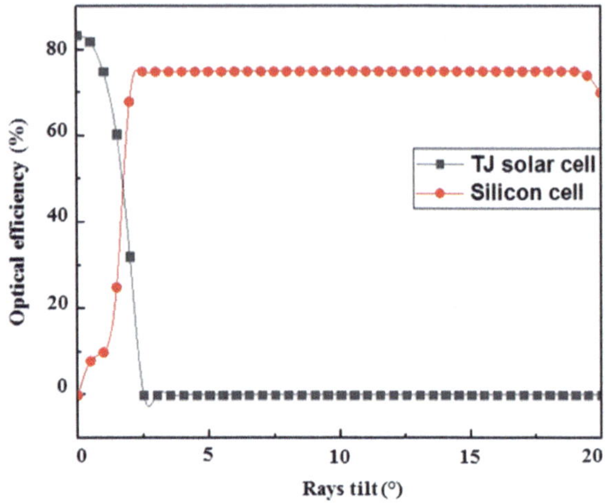

Fig. 10 Optical efficiency of the Hybrid micro-CPV

Table 6 Comparison results of optical efficiency

	Optical efficiency at TJ solar cell				Silicon solar cell			
	(%)		W		(%)		W	
	0°	20°	0°	20°	0°	20°	0°	20
Hybrid micro CPV	83.32	0	664	0	0	78.22	0	156

4 Conclusion

In this study we have covered a general, parametric and performance analysis of two micro CPV systems for car roof application using the same primary optic: the Fresnel lens (FL). It concerns a non-imaging optics which is: Pyramid, and an imaging optics: square dome.

The study starts from optical modeling of the Fresnel lens and considers several sizes and geometric needs of SOs. The performance of all concentrator systems is compared in terms of actual concentrations achieved, angles of acceptance, output light power distribution and tolerances in placement of the secondary element relative to the FL. The results are compared with the same types of SOs but are designed for car roof applications.

For the first simulation results, we found that the Fresnel lens with 28 prisms and a focal length of 14 mm, is the optimal primary optical element, and the square dome is the most suitable element as a secondary optical element placed at focal. By adding the silicon solar cell to multijunction solar cell, the efficiency of the optical system is increased by 20%. Therefore the power generated is 205 KW therefore this car can drive with a speed of 120 km/h.

References

1. Mkahl, R. Nait Sidi Moh, A., Wack, M.: Modeling and simulation of standalone photovoltaic charging stations for electric vehicles. In: XIII International Conference on Computer, Communication
2. Sanna, J.C., Gawronska, M., Damiano, A.: Energy performance of concentrator and flat plate photovoltaics in the mediterranean area. In: 2015 IEEE International Conference on Renewable Energy Research and Applications (ICRERA), Palermo, Italy, pp. 760–765, 22–25 Nov. 2015
3. Steiner, M., Siefer, G., Schmidt, T., Wiesenfarth, M., Dimroth, F., Bett, A.W.: 43 % Sunlight to electricity conversion efficiency using CPV. IEEE J. Photovolt. **6**(4), 1020–1024 (2016)
4. El Himer, S., Ahaitouf, A., El-yahyaoui, S., Mechaqrane, A.: Performance analysis of non-imaging Fresnel lens as a primary stage for CPV units. In: IOP Conference Series: Earth and Environmental Science, p. 012029 (2018). https://doi.org/10.1088/1755-1315/161/1/012029
5. Leutz, R., Suzuki, A., Akisawa, A., Kashiwagi, T.: Design of a nonimaging Fresnel lens for solar concentrators. Solar Energy **65**, 379–387 (1999). https://doi.org/10.1016/b978-008043895-5/50228-9
6. Akisawa, M., Hiramatsu, K.: Ozaki, Design of dome-shaped non-imaging Fresnel lenses 3 taking chromatic aberration into account. Sol. Energy **86**, 877–885 (2012). https://doi.org/10.1016/j.solener.2011.12.017
7. Pham, T.T., Vu, N.H., Shin, S.: Design of curved fresnel lens with high performance creating competitive price concentrator photovoltaic, Energy Procedia **144**, 16–32 (2018). https://doi.org/10.1016/j.egypro.2018.06.004
8. Languy, F., Lenaerts, C., Loicq, J., Thibert, T., Habraken, S.: Performance of solar concentrator made of an achromatic Fresnel doublet measured with a continuous solar simulator and comparison with a singlet, Sol. Energy Mater. Sol. Cells 70–76 (2013) . https://doi.org/10.1016/j.solmat.2012.10.008
9. Vallerotto, G., Victoria, M., Askins, S., Antón, I., Sala, G.: Experimental characterization of achromatic doublet on glass (ADG) Fresnel lenses, AIP Conf. Proc. **1881**, 030010 (2017). https://doi.org/10.1063/1.5001421
10. Zhuang, Z., Yu, F.: Optimization design of hybrid Fresnel-based concentrator for generating uniformity irradiance with the broad solar spectrum. Opt. Laser Technol. **60**, 27 (2014). https://doi.org/10.1016/j.optlastec.2013.12.021
11. Wang, G., Chen, Z., Hu, P., Cheng, X.F.: Design and optical analysis of the band-focus Fresnel lens solar concentrator. Appl. Therm. Eng. **102**, 695–700 (2016). https://doi.org/10.1016/j.applthermaleng.2016.04.030
12. Ryu, K., Rhee, J.G., Parkc, K.M., Kim, J.: Concept and design of modular Fresnel lenses for concentration solar PV system. Sol. Energy **80**, 1580–1587 (2006). https://doi.org/10.1016/j.solener.2005.12.006
13. Pham, T.T., Vu, N.H., Shin, S.: Novel design of primary optical elements based on a linear fresnel lens for concentrator photovoltaic technology. Energies **12**, 1209 (2019). https://doi.org/10.3390/en12071209
14. Chong, K.K., Siaw, F.L., Wong, C.W., Wong, G.S.: Design and construction of non-imaging planar concentrator for 7 concentrator photovoltaic system. Renew. Energy **34**, 1364–1370 (2009). https://doi.org/10.1016/j.renene.2008.09.001
15. Vallerotto, G., Victoria, M., Askins, S., Antón, I., Sala, G.: Experimental characterization of achromatic doublet on glass (ADG) Fresnel lenses. AIP Conf. Proc. **1881**, 030010 (2017). https://doi.org/10.1063/1.5001421
16. Lokeswaran, S., Mallick, T.K., Reddy, K.S.: Design and analysis of dense array CPV receiver for square parabolic dish system with CPC array as secondary concentrator. Sol. Energy **199**, 782–795 (2020). https://doi.org/10.1016/j.solener.2020.02.075
17. Tian, M., Su, Y.H., Fei Zheng, H., Pei, G., Li, G., Riffat, S.: A review on the recent research progress in the compound parabolic concentrator (CPC) for solar energy applications. Renew. Sustain. Energy Rev. **82**, 1272–1296 (2018). https://doi.org/10.1016/j.rser.2017.09.050

18. Yu, Y., Liu, N., Li, G., Tang, R.: Performance comparison of CPCs with and without exit angle restriction for concentrating radiation on solar cells, angle restriction for concentrating radiation on solar cells. Appl. Energy **155**, 284–293 (2015). https://doi.org/10.1016/j.apenergy.2015.06.012
19. Rodríguez, J.P.F., Baig, H., Fernández, E.F., Almonacid, F., Mallick, T., Higueras, P.P.: Optical modeling of four Fresnel-based high-CPV units. Sol. Energy **155**, 805–815 (2017). https://doi.org/10.1016/j.solener.2017.07.027
20. Rodríguez, J.P.F., Fernández, E.F., Baig, H., Almonacid, F., Mallick, T., Higueras, P.P.: Development, indoor characterisation and comparison to optical modelling of four Fresnel-based high-CPV units equipped with refractive secondary optics. Sol. Energy Mater. Sol. Cells **186**, 273–283 (2018). https://doi.org/10.1016/j.solmat.2018.06.050
21. El Himer, S., Ahaitouf, A., El-yahyaoui, S., Mechaqrane, A.: Parametric optimization and performances analysis of four secondary optical elements for concentrator photovoltaic systems. Int. J. Renew. Energy Res.-IJRER **8**, 4 (2018). https://doi.org/10.1063/1.5053502
22. Trespidi, F., Timò, G.: Optical Design of an off-axis mirror based solar concentrator. AIP Conf. Proceed. **1881**, 030008 (2017). https://doi.org/10.1063/1.5001419
23. Trespidi, F., Malvisi, E., Parmesani, R.: Shape characterization of the primary mirror of a mirror based solar concentrator. AIP Conf. Proceed. **1881**, 030009 (2017). https://doi.org/10.1063/1.5001420
24. Cruz-Campa, J.L., et al.: Microlens rapid prototyping technique with capability for wide variation in lens diameter and focal length. Microelectron. Eng. **87**(11), 2376–2381 (2010)
25. Yoon, J., et al.: Ultrathin silicon solar microcells for semitransparent, mechanically flexible and microconcentrator module designs. Nat. Mater. **7**(11), 907–915 (2008)
26. Domínguez, C., et al.: A review of the promises and challenges of micro-concentrator photovoltaics. AIP Conf. Proc. **1881** (2017)
27. Jared, B.H., et al.: Micro-concentrators for a microsystems-enabled photovoltaic system. Opt. Soc. Am. **22** (2014)
28. Ritou, A.: Développement, fabrication et caractérisation de modules photovoltaïques à concentration à ultra haut rendement à base de micro-concentrateurs. PhD thesis Grenoble Alpes University (2016)
29. Andreev, V.M., et al.: High-efficiency (24.6% AM 0) LPE grown AlGaAs/GaAs concentrator solar cells and modules. In Proceedings of the 1994 IEEE 1st World Conference on Photovoltaic Energy Conversion-WCPEC (A Joint Conference of PVSC, PVSEC and PSEC), Waikoloa, HI, USA, 5–9 December (1994).
30. Hayashi, N., et al.: Thin concentrator photovoltaic module with micro-solar cells which are mounted by self-align method using surface tension of melted solder. AIP Conf. Proc. **1881** (2017)
31. Ayane, S.E.L., Himer, S.E., El-Yahyaoui, S., Ahaitouf, A.: Performance analysis of a two stage micro photovoltaic concentrator. In: Proceedings of 2019 7th International Renewable and Sustainable Energy Conference, IRSEC 2019 (2019)

Machine Learning-Based Maximum Power Point Tracking Technique for Concentrated PV/TEG System Under Non-uniform Environmental Conditions

Noman Mujeeb Khan, Majad Mansoor, Syed Kumayl Raza Moosavi, Kamran Khan, Zakria Qadir, and Muhammad Hamza Zafar

Abstract This research work focuses on the hybrid concentrated Photovoltaic-Thermo electric (PV-TEG) module. The electrical characteristics of this framework give rise to complex control problems which can be classified as nonlinear, multiple local maxima monotonic control problems. Due to the existence of several gradient peak points, the conventional controllers cannot distinguish among the local and global maximum power points. Yet another problem with large-scale concentrated PV-TEG systems connected to the grid is the undesired large fluctuations in electrical power generation. Such surges and drops in voltage imbalances the grid oscillation balance and may cause failure and safety issues. This new approach is the first step toward stable power generation. For PV-TEG system power optimization, a hybrid Chaos Game Optimization Algorithm-General Neural Regression Network (CGOA-GRNN) is devised. A comprehensive comparison is made for performance evaluation with Henry gas optimization (HGO), Grey wolf optimization (GWO), and Particle Swarm Optimization (PSO). The statistical, qualitative, and quantitative results show superior performance by the proposed CGO-GRNN for concentrated solar thermal control application. The experimental simulation results indicate that the efficiency of the CGO-GRNN reaches up to 35.33%, TEG achieves 0.6% added efficiency, while proposed MPPT control achieves 100% GMPP detection score, with 99.9%

N. Mujeeb Khan · M. Hamza Zafar (✉)
Department of Electrical Engineering, Capital University of Science and Technology, Islamabad 44000, ISB, Pakistan
e-mail: hamzauetms@gmail.com

M. Mansoor
Department of Automation, University of Science and Technology of China, Hefei, China

S. Kumayl Raza Moosavi
National University of Sciences and Technology, Islamabad 44000, ISB, Pakistan

K. Khan
Department of Electrical Engineering, Hamdard University, Islamabad, Pakistan

Z. Qadir
School of Computing Engineering and Mathematics, Western Sydney University, Locked Bag 1797, Penrith, NSW 2751, Australia

S. El Himer et al. (eds.), *Artificial Intelligence of Things for Smart Green Energy Management*, Studies in Systems, Decision and Control 446,
https://doi.org/10.1007/978-3-031-04851-7_6

power tracking efficiency. The tracking time under fast varying solar concentration is on average 80 ms, while under partial shading (PS) conditions it is up to 90–110 ms. Additionally, the statistical analysis shows the robustness of the proposed methodology achieving a very low measure of variability.

Keywords Concentrated PV/TEG · Non-uniform conditions · Chaos game optimization · Maximum power point tracking · Statistical analysis.

1 Introduction

In recent times, the rising concern for global warming, availability of distributive renewable energy generation systems, and cost parity of domestic electrical power generation systems is attracting the masses to utilize small and medium-scale renewable energy resources. Among these resources solar systems, wind, and biofuel cells stand apart due to their positive impact on climate change problems. Renewable energy systems powered by solar energy are faced with low energy conversion efficiency. In the recent decade, a new tactic for the optimization of concentrated solar energy using the PV-TEG with commercially available solar cells and TEG is coming into the highlight. Because it enhances the electrical efficiency by 20–32% for existing setups. The efficiency of the TEG is directly proportional to the solar irradiation concentration. The solar irradiance of 330 W/m^2 may enhance output power by 3.5% on average.

The multi-junction solar cell (MJSC) converts a smaller portion of the incident solar radiation into electricity. The rest of the solar energy as heat is absorbed by the crystal lattice [1]. The rise in operating temperature occurs which lowers the open-circuit voltage of the PV system. The output power of PV module drops by 0.45% per °C rise in temperature. In the study of physical and thermal properties of TEG modules, it is estimated that for 150–250 K difference in temperature among the hot side and the cold side of the TEG leads to 4–6% conversion efficiency. The concept of combining a photovoltaic with a thermoelectric cooler or a thermoelectric generator using active combined MPPT control is a proposed model for the improvement of the overall system efficiency. Theoretically, the usage of the hybrid PV-TEG not only increases the efficiency but also improves the lifetime of a module by the cooling effect. A standard setup of concentrated solar PV-TEG system is shown in Fig. 1.

PV systems are the most economical and extensively investigated renewable energy resource. PV systems convert solar energy into the most effective electrical energy. The leaps in manufacturing techniques, advancement in material and favorable legislation are the main contributors to the explosive growth of PV systems. The least carbon footprint, ease of installation, and scalability are the main qualities for building-integrated PV systems (BIPV). The applications of PV range from floating PV systems (FPV), commercial solar power plants, wearable electronics, light poles, microgrids, and charging stations [2].

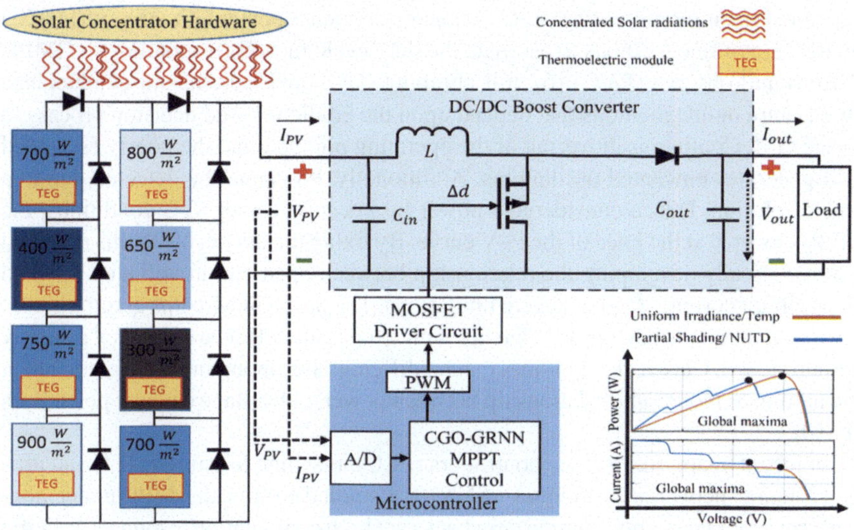

Fig. 1 A standard hybrid PV-TEG system block diagram with solar irradiation concentrator, PV-TEG network, control and sensor block and load interface blocks

The MPPT is a challenging problem for optimization control. The first and far most reason is that the PV model electrical characteristics are nonlinear. Secondly, the operating point is sensitive to operating conditions such as irradiance (G) and temperature (T) [3]. The third problem is caused by the irregular distribution of irradiance across the PV panels connected in a series–parallel configuration. This is known as partial shading (PS) and is a major cause for failure of attaining maximum power by conventional MPPT techniques. The main requirement from the MPPT technique for PV systems as a utility resource is fast-tracking of MPP, high efficiency, and zero oscillations in voltage and current transients [4]. The MPPT controllers provide an interface between power generation and load. It is done by the use of DC converters and inverters [5]. The PV system is operated on MPP by fine-tuning the duty cycle of the boost converter. The change in irradiance and temperature level shifts the GMPP. Therefore, it is necessary to actively track the MPP. Figure 1 presents the components of a standard PV system such as load, controller, and PV-TEG arrays.

1.1 Literature Review

In literature, the application of artificial intelligence techniques in renewable energy has enormous potential. The cost–benefit analysis of MPPT control favors low-cost implementation. Many studies have been carried out to find efficient methods of passive and active MPPT. The main categories of MPPT controllers are offline and online which overlap between classical and intelligent controllers. Pilot cell, fractional open-circuit voltage (FOCV) and fractional open circuit current (FOCC) models are offline methods to estimate the duty cycle ratio corresponding to GMPP. Perturb and observe (P&O) [6], hill climbing (HC), and incremental conductance (IC) [7] are online methods that depend upon the gradient-based decision process to locate GM. Continuous hovering of the operating point around the poorly estimated GM generates unwanted oscillations. Additionally, PS cannot be detected due to a single solution. Hence considerable power loss occurs. Under STC conditions, the MPP is located at the knee of the P–V curve. By fixing the duty cycle to this ratio and mathematically developing the relationship between open-circuit voltage (V_{0c}) and short circuit current (I_{sc}) has severe limitations. The problem with this assumption of the constant ratio is that the V_{oc} changes with temperature. Hence efficiency is highly compromised. Classical techniques cannot differentiate among multiple gradients in the nonlinear monotonic relationship between power and voltage causing power loss [8–10].

In recent years, fuzzy logic controllers (FLC), machine learning (ML), and artificial neural networks techniques (ANN) are studied to find the MPP under non-uniform irradiance. The common drawbacks are high computation resources, training data sets, algorithm complexity, expensive micro-controllers, and lack of standard analysis to gauge relative performance [11].

MPPT can be regarded as an optimization problem. Meta-heuristic evolutionary algorithms specifically swarm intelligence-based optimization algorithms provide huge potential to address the limitations of FLC, ML, and classical MPPT techniques [12]. The SI can well handle the nonlinear multiplicity of the base system such as PV systems with less accurate equivalent modeling. The studied CGO algorithm is an extraordinary member of optimization algorithms and has the potential to track GM under PS. In this class particle swarm optimization (PSO) [13, 14], grasshopper optimization (GHO) [15], artificial bee colony (ABC) [16], simulated annealing (SA) [17], cuckoo search (CS) [18, 19], overall distribution (OD) algorithm [20], ant colony optimization (ACO) [21, 22], adaptive cuckoo optimization algorithm (ACOA) [23], farmland fertility optimization (FFO) [24], Harris hawk optimization (HHO) [25], salp swarm optimization algorithm (SSA) [26], grey wolf optimization (GWO) [27], spline guided Jaya (SGJ) algorithm[28], Butterfly optimization algorithm (BOA) [29], and particle swarm optimization gravitational search (PSOGS) have been studied. Performance of SI is subjected to factors such as population size, information sharing mechanism, epoch iterations, convergence speed, parametric tuning, and computational load [30].

The rest of the study is divided into five sections. Section 1 presents the problem formulation and electrical characteristics of the PV-TEG system. Section 2 introduces MPPT based CGO-GRNN technique. In Sect. 3, the results are analyzed. Section 4 gives the summary and Sect. 5 provides the concluding remarks of the research results.

2 Modelling of PV Cell and TEG Module

2.1 PV Cell Model

Single diode model (SDM) equivalent electrical circuit of a PV device illustrated in Fig. 2. The current flowing out of a PV cell can be expressed as:

$$I_T = I_o + I_{sh} + I_l \tag{1}$$

The mathematical model is given as:

$$I_T = I_P - I_s\left(e^{\frac{V+IR_s}{\alpha V_T}} - 1\right) - \frac{(V + I_T Rs)}{R_{st}} \tag{2}$$

$$I_P = (I_{sc-STC} + k_i \nabla T)\frac{\mathrm{G}}{\mathrm{G}_N} \tag{3}$$

$$I_s = \frac{I_{sc}}{(e^{\frac{V_{oc}}{\alpha v_T}} - 1)} \tag{4}$$

where I_T is load current and V is the terminal voltage of the PV cell,I_P is the current produced by irradiance G, I_s is the diode saturation current, k_i is temperature coefficient for short circuit current, G_N is irradiance at STC. R_{st} and R_{st} are the parallel and series resistances. $v_T$is he the thermal voltage constants of diode. Depending on the number of cells linked in series and parallel (Ns and Np respectively), SDM may be scaled up or down to form a single PV cell, a PV module, or a PV array.

$$I_l = n_p I_P - n_p I_s\left(e^{\frac{V+(\frac{n_s}{n_p})IR_s}{\alpha n_s V_T}} - 1\right) - \frac{(V + (\frac{n_s}{n_p})I_l Rs)}{(\frac{n_s}{n_p})R_{st}} \tag{5}$$

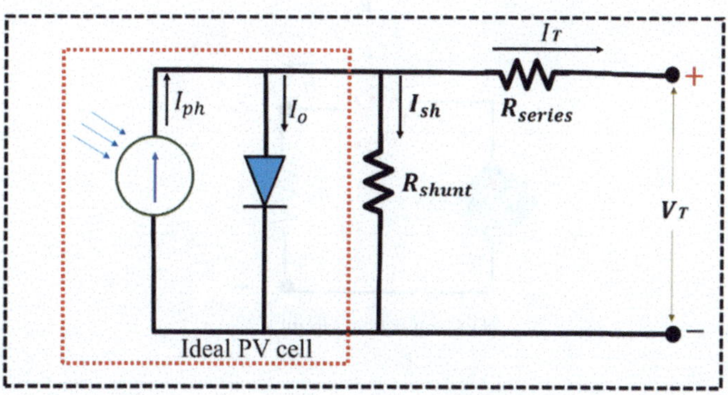

Fig. 2 PV electrical model

where n_p and n_s are number of cells connected in series and parallel.

2.2 TEG Model

Figure 3 represents basic structure and equivalent electrical model of thermoelectric module. Thermoelectric module consists of thermocouples connected in series. Electrically thermoelectric generator can be represented by a voltage source with an internal resistance. Assuming a constant temperature differential between junctions, open circuit voltage V_{oc} of a thermoelectric module may be theoretically represented as

$$V_{oc} = \alpha_{p-n}.\Delta T = \alpha_{p-n}(T_h - T_c) \quad (6)$$

where α_{p-n} is symbolic representation of Seebeck coefficient. ΔT is temperature difference between hot side (T_h) and cold side (T_c) junctions.

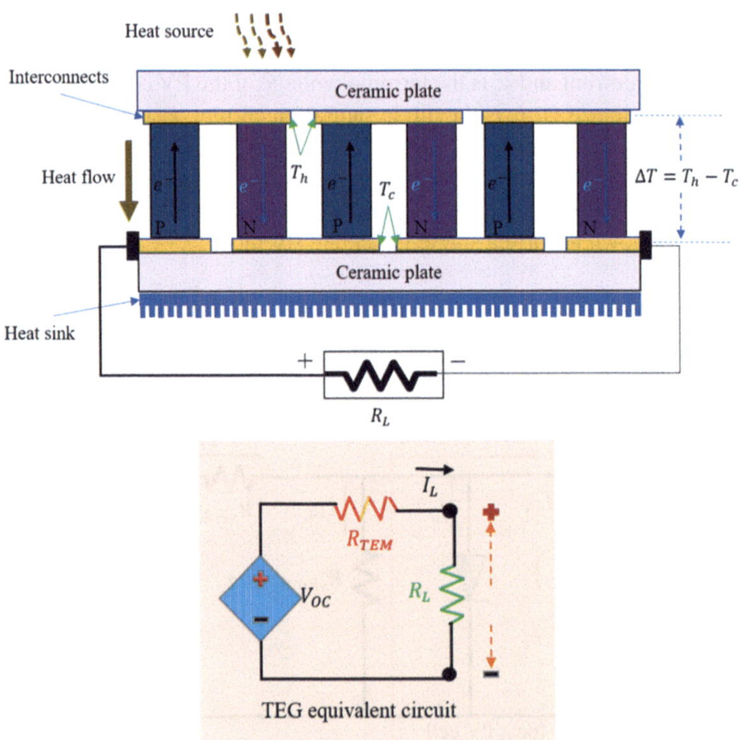

Fig. 3 TEG structure and equivalent electrical model

The TEG converts thermal energy to electrical energy based on Seebeck effect, a phenomenon in which temperature gradient between two different metals or semi-conductors induce e.m.f between them. Seebeck coefficient depends on the material of TEG and gives measure of the voltage induced in response to temperature difference. Additionally, TEG is also affected by Thomson effect which can be modelled as:

$$\tau = T\frac{d\alpha_{p-n}}{dT} \tag{7}$$

where τ represents Thomson coefficient and T is the average temperature. Due to Thomson effect, Seebeck coefficient becomes a function of average temperature and will dynamically varies with change of average temperature. This behavior can be modelled as:

$$\alpha(T) = \alpha_o + \alpha_1 \ln(T/T_o) \tag{8}$$

where α_o represents basic Seebeck coefficient and α_1 is the variation rate of Seebeck coefficient due to Thomson effect at a reference temperature T_o. From the equivalent circuit load current is a function of load resistance and internal resistance of thermoelectric module R_{TEM} and can be written as,

$$I_L = \frac{V_{oc}}{R_L + R_{TEM}} \tag{9}$$

Hence power generated by TEG can be written as,

$$P_{TEG} = (\alpha_{p-n}.\Delta T)^2.\frac{R_L}{(R_L + R_{TEM})^2} \tag{10}$$

2.3 Concentrated PV/TEG Module

Generally, the PV cells utilize only a partial amount of incident solar irradiation for generating useful electricity, and the larger amount of irradiation is converted into waste heat energy. Furthermore, the waste heat energy generated on the PV panel raises the cell temperature, lowering the PV system's energy efficiency. The TEG with heat sinks is installed in the proposed system to utilize waste heat energy for increased power generation while also lowering the PV cell temperature. A typical structure of the PV and TEG combined hybrid module is depicted in Fig. 4. The overall system power conversion efficiency is expressed as follow:

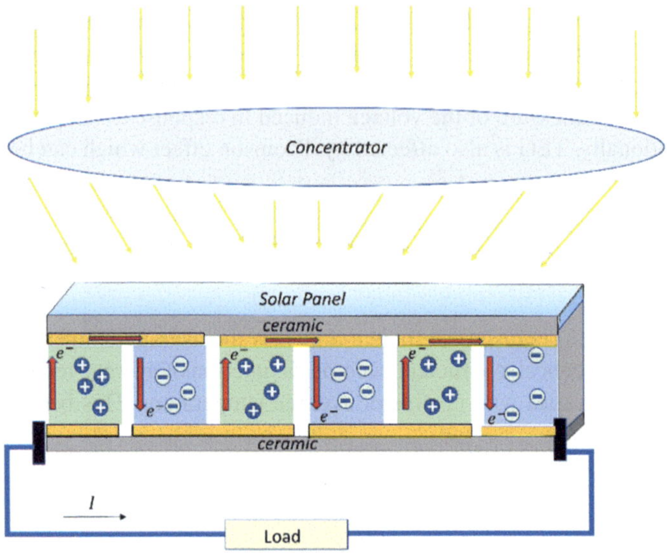

Fig. 4 The physical structure of hybrid PV-TEG system

$$\eta_{pv-teg} = \frac{P_p + P_{TEG}}{G \times A_p} \quad (11)$$

where P_p is PV panel output power, P_{TEG} is TEG output power and A_p represents area of PV panel.

Multiple PV-TEG module combinations exist to generate high power for various purposes. Figure 5. depicts the centralized PV-TEG system. Use of single DC converter for control and power distribution to load results in cost reduction. However, stochastic environmental conditions may result in many local maximum power points, making maximum power tracking task more complex. Therefore, this work provides an effective control technique for hybrid photovoltaic thermoelectric generation (HPVTEG) systems to extract maximum power under NTD and partial shading.

2.4 *Non-uniform Conditions*

In concentrate PV/TEG system elements are linked to a single boost converter. In a dynamic environment it is not necessary that all connected elements operate under same temperature and irradiance conditions. The PV/TEG module with lowest irradiance and ΔT limits performance of a PV/TEG string. The least performing module acts as heat sink, resulting in hot spot formation and may cause permanent damage

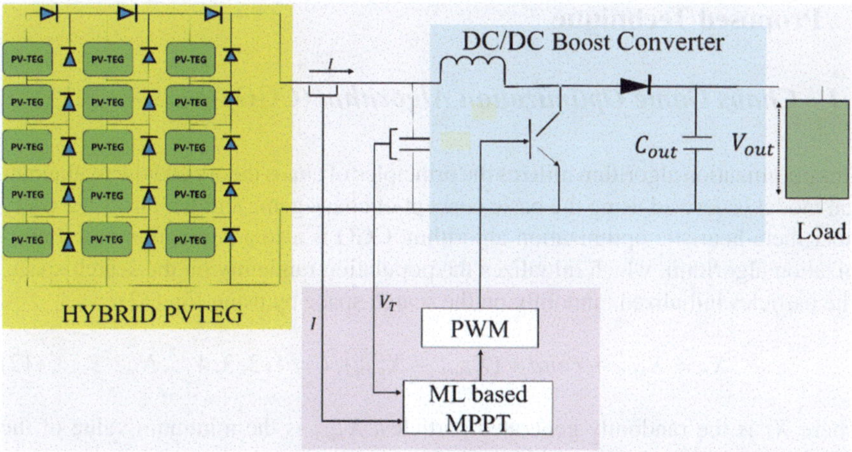

Fig. 5 The CGO-based MPPT for HPVTEG module

to PV/TEG string. The placement of a bypass diode across each module is a straightforward approach for isolating the least productive string. However, use of bypass diodes results in several peaks on the power voltage curve of PV/TEG module, as seen in Fig. 6b. There is just one peak under uniform temperature circumstances, as seen in Fig. 6a. Therefore, using an MPPT approach to maximize the energy harvest from PV/TEGs is critical for PV/TEG systems operating under NUC conditions.

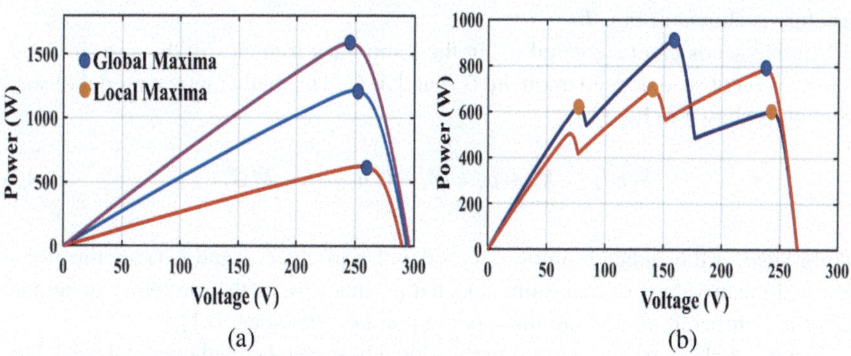

Fig. 6 **a** Power voltage (PV) curve of PV/TEG under uniform temperature distribution, **b** PV curve of PV/TEG under non-uniform temperature distribution

3 Proposed Technique

3.1 *Chaos Game Optimization Algorithm (CGO)*

This optimization algorithm utilizes the principles of chaos theory and the mathematical model is derived using the basic concept of chaos game and fractals. Like every other meta-heuristic optimization algorithm, CGO is also a population based optimization algorithm, which initializes the population randomly on the search space. The particles initialized randomly on the search space by using Eq. (12).

$$X_i = X_{min} + rand * (X_{max} - X_{min}), i = 1, 2, 3, 4 \ldots N \tag{12}$$

where X_i is the randomly generated particles, X_{min} is the minimum value of the search space, X_{max} is the maximum value of search space, $rand$ is the random number generated between 0 and 1.

The main idea formulated for the mathematical modelling of this algorithm is to complete the Sierpinski triangle by creating different seeds in the search space. The temporary triangle is created using three seeds, which are as follows:

- Global Best (GB): Best solution evaluated up to the current iteration
- Mean Best (MG): The position of mean group.
- Current Solution (X_i): The position of the current solution

The GB and MG combined with X_i makes the three points of the Sierpinski triangle. The view of creating the temporary triangles is shown in Fig. 7a and detailed structure is shown in Fig. 7b.

Eligible seeds can be created using the temporary triangles in the search space. The creation of seed depend upon the GB and MG. The mathematical model of seed creation is shown in Eq. (13).

$$Seed_i = X_i + \alpha_i * (\beta_i * GB - \gamma_i * MG_i) \tag{13}$$

where $Seed_i$ is the updated solution, X_i is the ith particle, GB and MG are the global best and mean values of randomly selected particles, α_i is the randomly generated factorial number, β_i and γ_i are the random numbers between [0,1].

Second seed can be updated using the global best and the mathematical model for this is shown in Eq. (14).

$$Seed_i = GB + \alpha_i * (\beta_i * GB - \gamma_i * MG_i) \tag{14}$$

where $Seed_i$ is the updated solution, X_i is the ith particle, GB and MG are the global best and mean values of randomly selected particles, α_i is the randomly generated factorial number, β_i and γ_i are the random numbers between [0,1].

Since MG is the third seed which can be used to create new seed by using Eq. (15).

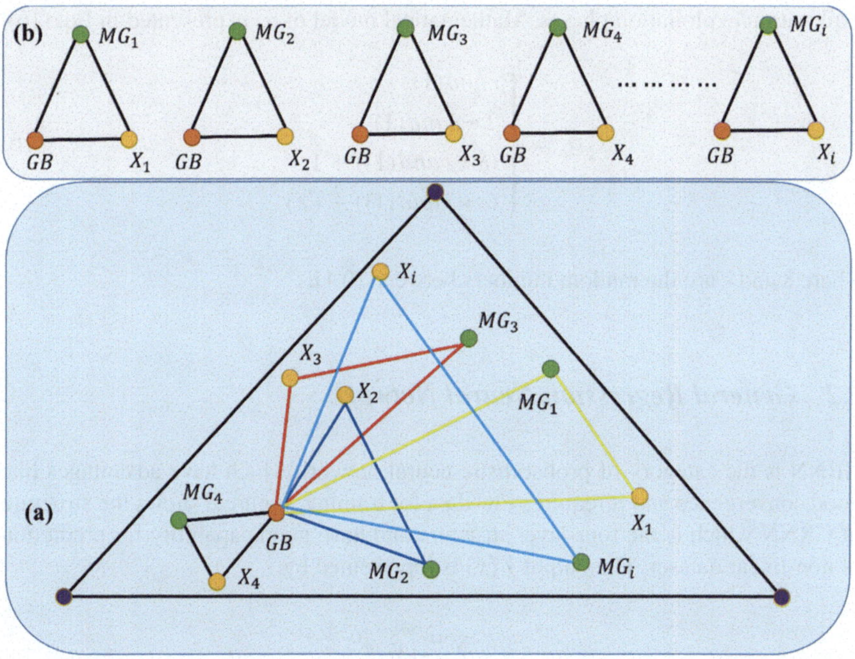

Fig. 7 CGO Working Mechanism

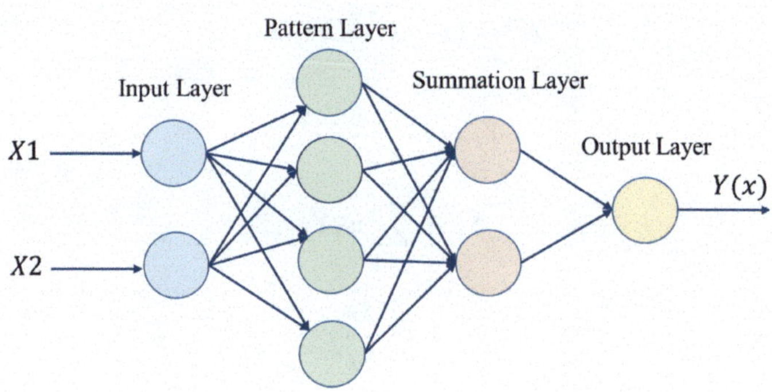

Fig. 8 Structure of GRNN

$$Seed_i = MG + \alpha_i * (\beta_i * GB - \gamma_i * MG_i) \tag{15}$$

where $Seed_i$ is the updated solution, X_i is the ith particle, GB and MG are the global best and mean values of randomly selected particles, α_i is the randomly generated factorial number, β_i and γ_i are the random numbers between [0,1].

Exploration and exploitation are the important phases in metaheuristic optimization algorithms. In CGO α_i is the controlling parameter, which controls the exploration/exploitation phases. Mathematical model of α_i is presented in Eq. (16).

$$\alpha_i = \begin{cases} rand(1) \\ 2 * rand(1) \\ (\delta * rand(1)) + 1 \\ (\varepsilon * rand(1)) + ('\varepsilon) \end{cases} \tag{16}$$

where δ and ε are the random numbers between [0,1].

3.2 General Regression Neural Network

GRNN is the category of probabilistic neural network which have advantages like good convergence and it requires low data for training. Figure 8 shows the structure of GRNN which is the four-layer structure and have good capability for prediction of non-linear dataset. The output $Y(x)$ is represented by:

$$Y(x) = \frac{\sum_{j=1}^{m} w_j e^{(-\frac{K_j^2}{2\sigma^2})}}{\sum_{j=1}^{m} e^{(-\frac{K_j^2}{2\sigma^2})}} \tag{17}$$

$$K_j^2 = (x - x_j)^T (x - x_j) \tag{18}$$

$$Y_j = e^{-\frac{(x-x_j)^T (x-x_j)}{2\sigma^2}} \tag{19}$$

$$S_w = \sum_{j=1} w_j Y_j \tag{20}$$

$$Y = \frac{S_\delta}{S_w} \tag{21}$$

3.3 Neural Network Training Using a Bio-Inspired Algorithm for MPPT Control

Meta-heuristic algorithm is basically used to generate the dataset which is the voltage and current as input while the duty cycle as the output. The inputs are given at the input layer, while the pattern layer generates the pattern and summation layer do the sum of the output from the hidden layer. At the output layer, the final output

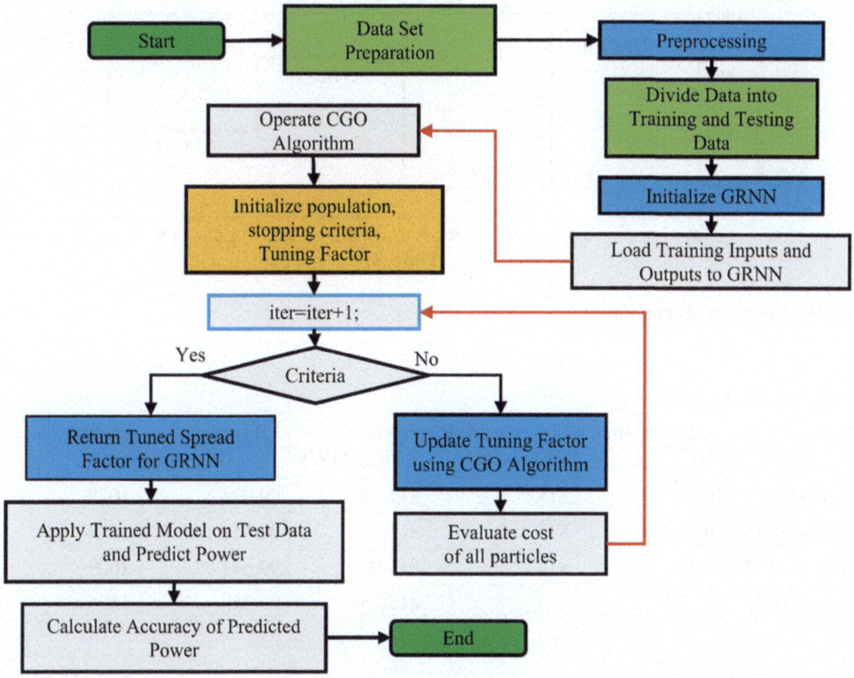

Fig. 9 Flow chart of GRNN training by CGO

is extracted. GRNN suffer from low prediction accuracy if the parameters, e.g. σ smoothing parameter with the number of neurons in GRNN do not tune well. For effective training and testing these parameters are needed to be optimally tuned. In this paper, CGO is used to tune these parameters, and MPPT control technique is implemented for concentrated PV/TEG system. The flow chart for the training of GRNN using CGO is shown in Fig. 9. Trained models are then tested on the test data and the performance evaluation is done.

4 Results and Discussion

4.1 Case 1: Varying Conditions

For testing ability of algorithm to adjust with changing environmental conditions step variation of irradiance is considered in case1. The irradiance varies after every 0.5 s as shown in the Fig. 10. Table 1 delineates summary of results. The powers achieved by CGO-GRNN is 817.25 W, HGO-GRNN is 816.82 W, GWO-GRNN is 815.07 W

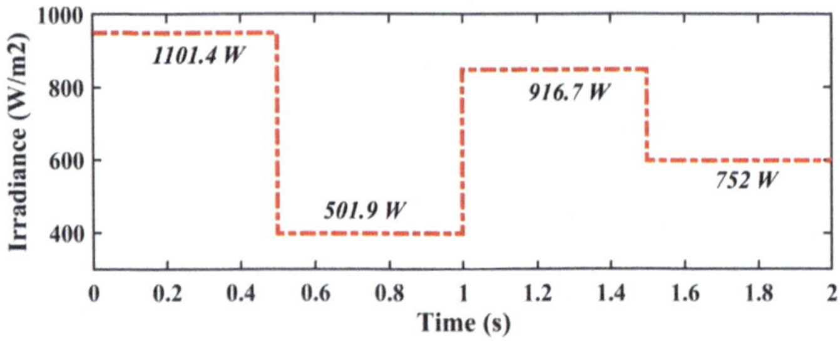

Fig. 10 Case 1: test scenario

Table 1 Case 1 results

Tech	Tracking time (s)	Power at MPP (W)	Power achieved (W)	Efficiency (%)	Energy (W.s)
CGO-GRNN	0.080	818	817.25	99.978	1628
HGO-GRNN	0.120	818	816.82	99.855	1613
GWO-GRNN	0.110	818	815.07	99.641	1609
PSO-GRNN	0.150	818	812.75	99.358	1595

and PSO-GRNN is 812.75 W with highest tracking efficiency of 99.978%, by CGO-GRNN. The tracking time for CGO-GRNN is 0.080 s, followed by 0.120 s for HGO-GRNN, 0.110 for GWO-GRNN and 0.150 s for PSO-GRNN. Power and voltage variations are shown in figure. Figure verified that the energy attained by CGO-GRNN is maximum while that of PSO-GRNN is minimum. Due to fast response and perfect balance between exploration and exploitation phase the tracking times is smallest for CGO-GRNN and highest for PSO-GRNN as sown in Fig. 11. Results clearly validates superior performance of proposed CGO-GRNN technique under changing irradiance conditions. The duty cycle as control parameter in terms of

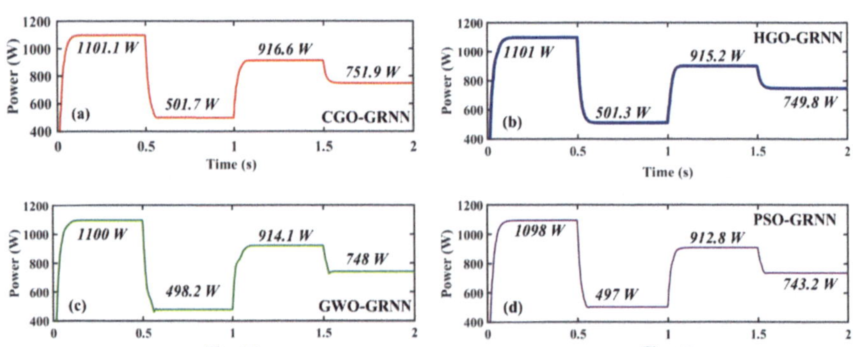

Fig. 11 Case 1: power comparison

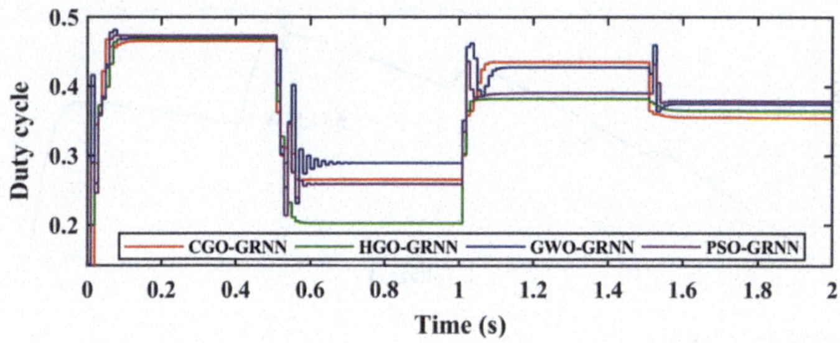

Fig. 12 Case 1: duty cycle comparison

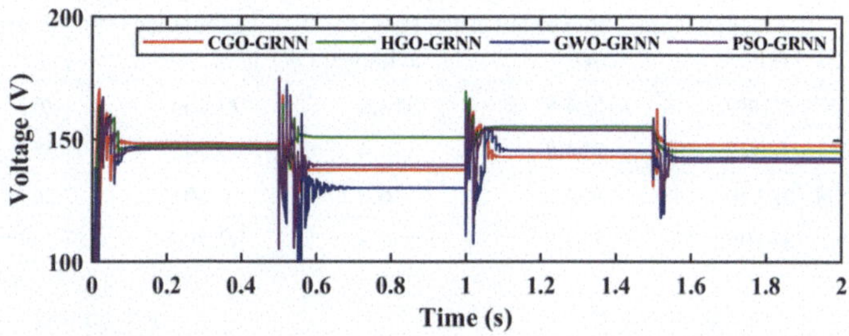

Fig. 13 Case 1: voltage comparison

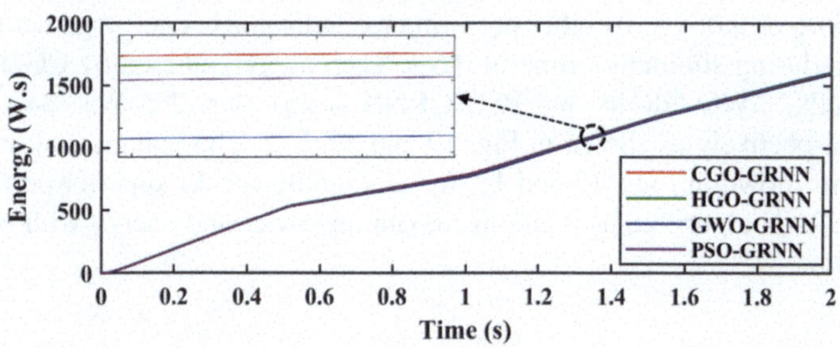

Fig. 14 Case 1: energy comparison

iteration is compared in Fig. 12. The corresponding voltages and currents transients of competing techniques are shown in Figs. 13 and 14 respectively.

4.2 Case 2: Non-Uniform Conditions

In case 2 due to stochastic environmental conditions multiple local maxima exists with a single global maximum at 554.8 W as shown in Fig. 15. Summary of results is given by Table 2. The ability of algorithm is judged based on how much and how

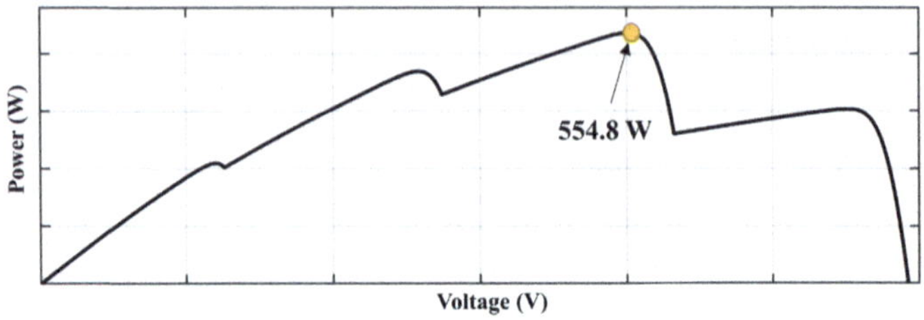

Fig. 15 Case 2: test scenario

Table 2 Case 2 results

Tech	Tracking time (s)	Power at MPP (W)	Power achieved (W)	Efficiency (%)	Energy (W.s.)
CGO-GRNN	0.089	554.8	554.6	99.963	261
HGO-GRNN	0.092	554.8	553.4	99.747	255
GWO-GRNN	0.110	554.8	536.3	96.665	247
PSO-GRNN	0.105	554.8	514.8	92.790	244

fast it tracks power. Powers attained by CGO-GRNN is 554.6 W then comes HGO-GRNN 553.4 W, GWO-GRNN 536.3 W and PSO-GRNN 514.8 W respectively as depicted in Fig. 16. CGO-GRNN shows highest efficiency of 99.963% with lowest tracking time of 0.089 s. Another performance indicator is energy gathered by the algorithms during simulation time of 0.5 s. The energy gathered by CGO-GRNN, HGO-GRNN, GWO-GRNN and PSO-GRNN is 261 W.s, 255 W.s, 247 W.s and 244 W.s respectively as shown in Fig. 19 and Table 2. The voltage and duty cycle changes are shown in Figs. 17 and 18. Results loudly speaks superior performance of CGO-GRNN algorithm as it attains maximum power and energy with minimum tracking time.

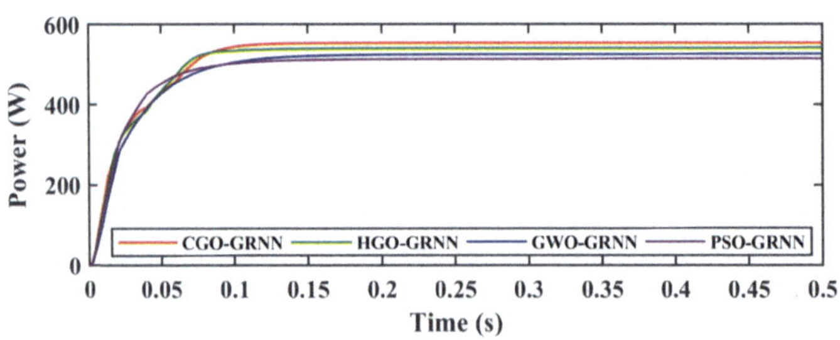

Fig. 16 Case 2: power comparison

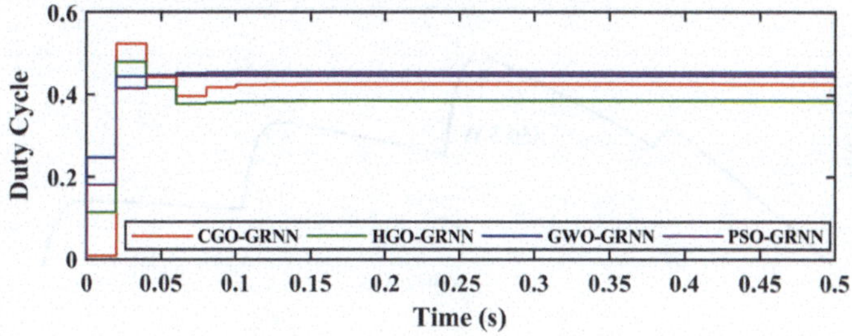

Fig. 17 Case 2: duty cycle comparison

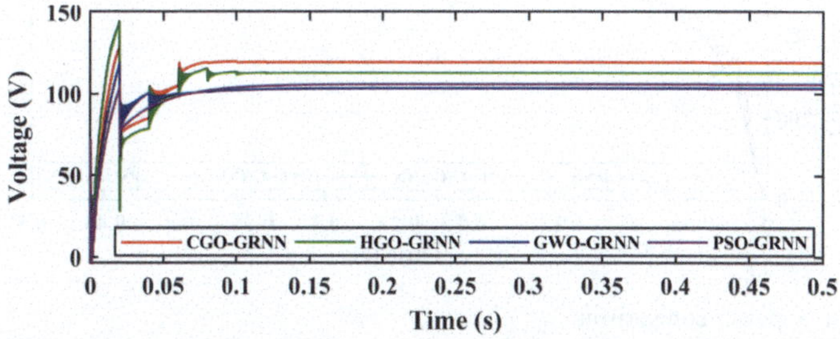

Fig. 18 Case 2: voltage comparison

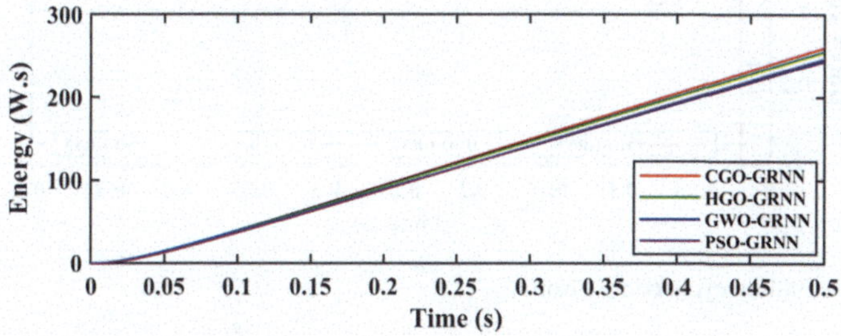

Fig. 19 Case 2: energy comparison

4.3 Case 3: Non-Uniform Conditions

This case is formulated to test performance of algorithm under left skewed GMPP conditioned. Figure 20 show that multiple local maxima exist with a single left skewed global maximum at 550.5 W. Figure 21 depicts results for Power transients, Fig. 22 control signal duty cycle, Fig. 23 voltage and energy accumulation is depicted in Fig. 24. Table 3 summarizes the results that validate superior performance of CGO-GRNN in terms of fast tracking, higher efficiency, power and energy. Slow converging speed oscillations in convergence stages of PSO results in lowest energy as compared to other algorithms.

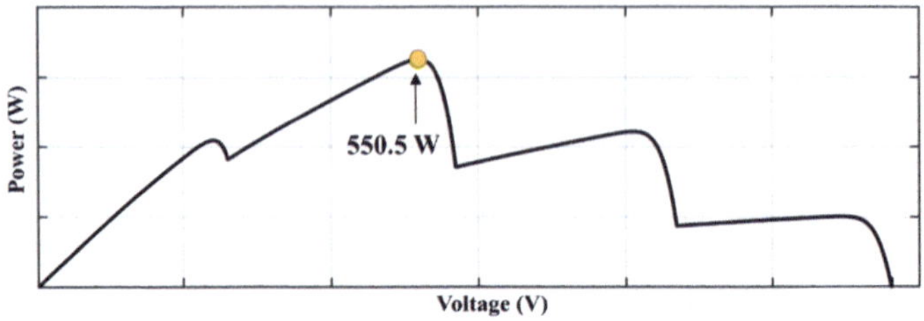

Fig. 20 Case 3: test scenario

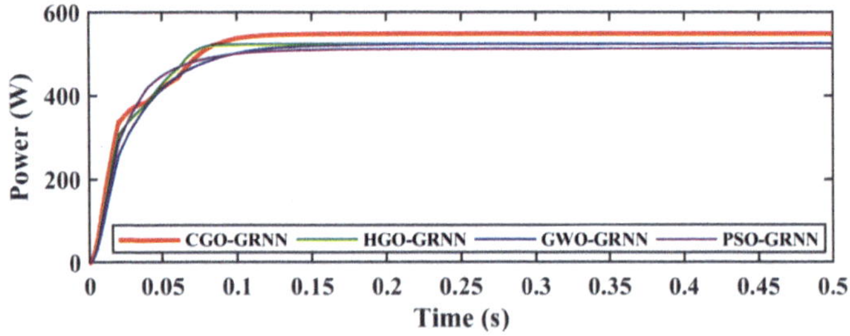

Fig. 21 Case 3: power comparison

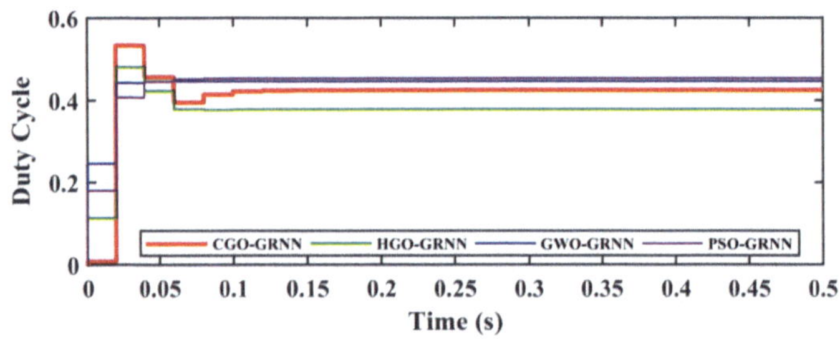

Fig. 22 Case 3: duty cycle comparison

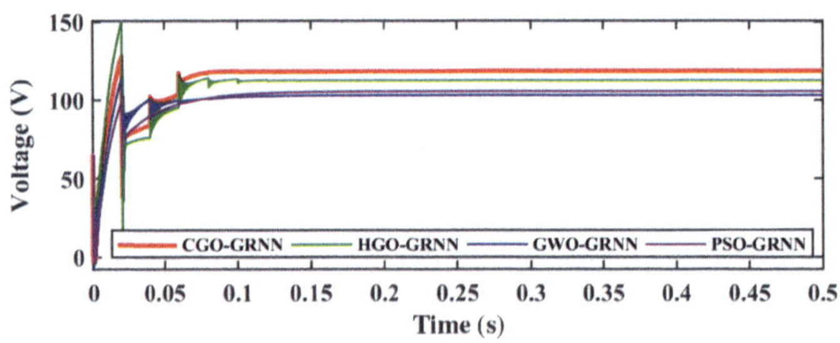

Fig. 23 Case 3: Voltage comparison

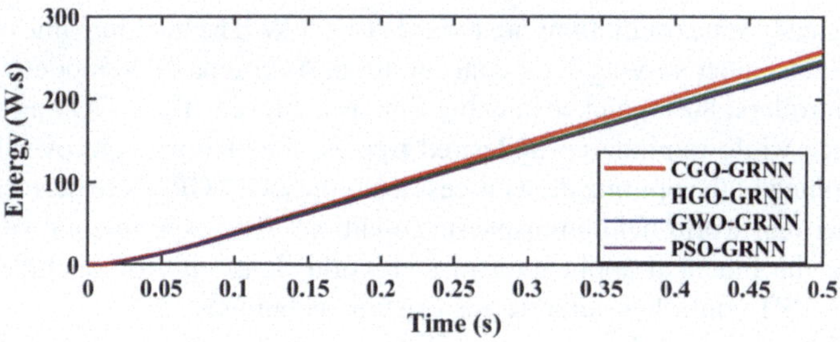

Fig. 24 Case 3: energy comparison

Table 3 Case 3 results

Tech	Tracking time (s)	Power at MPP (W)	Power achieved (W)	Efficiency (%)	Energy (W.s.)
CGO-GRNN	0.091	550.5	550.3	99.96	258
HGO-GRNN	0.097	550.5	541.5	98.36	251
GWO-GRNN	0.114	550.5	541.4	98.30	251
PSO-GRNN	0.109	550.5	520.3	94.51	239

4.4 Performance Evaluation

In the concentrated Solar-TEG, the TEG module acts as a cooling factor for the solar cells. This result shows improved performance of solar cells. The hybrid PV-TEG system proposed for concentrated PV applications in hot sites give good performances while the additional cost is about 15% of the total cost of classical PV systems. The Cost benefit analysis shows cost recovery within first years in arid environments with higher average sun exposure.

5 Conclusion

In this study, we have proposed intelligent MPPT control technique using a hybrid SI-NN framework for the concentrated solar-thermal power system. This article thoroughly examines the drawbacks of back solar thermal energy conversion system and provides solution to the technical limitations of classical gradient based MPPT control techniques. The research fills the gap of a thorough examination of the hybrid control technique while undertaking the problems of PS and slow convergence in MPPT control. The results are compared with highly optimized PHGO-GRNN, GWO-GRNN and PSO-GRNNs under various operating conditions. The statistical indices, analytical analysis and results comparison solidifies the superior performance of the proposed MPPT controller. The power tracking efficiency is up to 99.9%, steady state oscillations are reduced to ≤ 1 W. The tracking time of CGO is the least 80 ms under varying solar concentration. As compared to modern SI-based MPPT controllers, 80% quicker tracking time is achieved. The CGO has enhanced performance while minimizing undesired ripples, fluctuations, and overshoot. An effective tuning of the optimization process leads the CGO-GRNN to performs effectively under real world field atmospheric conditions. The experimental verification confirming the practical application in real-world. It is safe to conclude that the proposed MPPT controller outperforms existing techniques.

In the future, the MPPT problem will be extended to other renewable energy sources to enhance the productivity of the power generation by renewables. Higher efficiency operation enhances the economic feasibility of renewable sources encouraging the utility to combat global warming.

References

1. Kane, A.N., Verma, V.: Performance enhancement of building integrated photovoltaic module using thermoelectric cooling. Int. J. Renew. Energy Res. (IJRER) **3**(2), 320–324 (2013)
2. Fathabadi, H.: Novel stand-alone, completely autonomous and renewable energy based charging station for charging plug-in hybrid electric vehicles (PHEVs). Appl. Energy **260**, 114194 (2020)
3. Vicente, E.M., et al.: High-efficiency MPPT method based on irradiance and temperature measurements. IET Renew. Power Gener. **14**(6), 986–995 (2020)
4. Chen, Y., et al.: Performance and optimization of a novel solar-driven liquid desiccant air conditioning system suitable for extremely hot and humid climates. Energy Convers. Manage. **215**, 112899 (2020)
5. Li, S.: A variable-weather-parameter MPPT control strategy based on MPPT constraint conditions of PV system with inverter. Energy Convers. Manage. **197**, 111873 (2019)
6. Karabacak, M.: A new perturb and observe based higher order sliding mode MPPT control of wind turbines eliminating the rotor inertial effect. Renew. Energy **133**, 807–827 (2019)
7. Harrag, A., Messalti, S.: Ic-based variable step size neuro-fuzzy mppt improving pv system performances. Energy Procedia **157**, 362–374 (2019)
8. Ravyts, S., et al.: Embedded BIPV module-level DC/DC converters: classification of optimal ratings. Renew. Energy **146**, 880–889 (2020)
9. Pathak, D., Sagar, G., Gaur, P.: An application of intelligent non-linear discrete-PID controller for MPPT of PV system. Procedia Comput. Sci. **167**, 1574–1583 (2020)
10. Motahhir, S., El Hammoumi, A., El Ghzizal, A.: The most used MPPT algorithms: review and the suitable low-cost embedded board for each algorithm. J. Clean. Prod. **246**, 118983 (2020)
11. Chen, Z., et al.: Accurate modeling of photovoltaic modules using a 1-D deep residual network based on IV characteristics. Energy Convers. Manage. **186**, 168–187 (2019)
12. Bollipo, R.B., Mikkili, S., Bonthagorla, P.K.: Critical review on PV MPPT techniques: classical, intelligent and optimisation. IET Renew. Power Gener. **14**(9), 1433–1452 (2020)
13. Renaudineau, H., et al.: A PSO-based global MPPT technique for distributed PV power generation. IEEE Trans. Industr. Electron. **62**(2), 1047–1058 (2014)
14. Ishaque, K., et al.: An improved particle swarm optimization (PSO)–based MPPT for PV with reduced steady-state oscillation. IEEE Trans. Power Electron. **27**(8), 3627–3638 (2012)
15. Mansoor, M., et al.: Novel grass hopper optimization based MPPT of PV systems for complex partial shading conditions. Sol. Energy **198**, 499–518 (2020)
16. Soufyane Benyoucef, A. et al.: Artificial bee colony based algorithm for maximum power point tracking (MPPT) for PV systems operating under partial shaded conditions. Appl. Soft Comput. **32**, 38–48 (2015)

17. Lyden, S., Haque, M.E.: A simulated annealing global maximum power point tracking approach for PV modules under partial shading conditions. IEEE Trans. Power Electron. **31**(6), 4171–4181 (2015)
18. Ahmed, J., Salam, Z.: A maximum power point tracking (MPPT) for PV system using Cuckoo search with partial shading capability. Appl. Energy **119**, 118–130 (2014)
19. Ma, J., et al.: Parameter estimation of photovoltaic models via cuckoo search. J. Appl. Math. (2013)
20. Li, H., et al.: An overall distribution particle swarm optimization MPPT algorithm for photovoltaic system under partial shading. IEEE Trans. Industr. Electron. **66**(1), 265–275 (2018)
21. Jiang, L.L., Maskell, D.L., Patra, J.C.: A novel ant colony optimization-based maximum power point tracking for photovoltaic systems under partially shaded conditions. Energy Build. **58**, 227–236 (2013)
22. Kinattingal, S., Simon, S.P., Nayak, P.S.R.: MPPT in PV systems using ant colony optimisation with dwindling population. IET Renew. Power Gener. **14**(7), 1105–1112 (2020)
23. Mirza, A.F., et al.: Novel MPPT techniques for photovoltaic systems under uniform irradiance and partial shading. Sol. Energy **184**, 628–648 (2019)
24. Agwa, A.M., El-Fergany, A.A., Maksoud, H.A.: Electrical characterization of photovoltaic modules using farmland fertility optimizer. Energy Convers. Manage. **217**, 112990 (2020)
25. Mansoor, M., Mirza, A.F., Ling, Q.: Harris hawk optimization-based MPPT control for PV systems under partial shading conditions. J. Cleaner Prod. 122857 (2020)
26. Mirza, A.F., et al.: A salp-swarm optimization based MPPT technique for harvesting maximum energy from PV systems under partial shading conditions. Energy Convers. Manage. **209**, 112625 (2020)
27. Mohanty, S., Subudhi, B., Ray, P.K.: A new MPPT design using grey wolf optimization technique for photovoltaic system under partial shading conditions. IEEE Trans. Sustain. Energy **7**(1), 181–188 (2015)
28. Huang, C., et al.: A prediction model-guided Jaya algorithm for the PV system maximum power point tracking. IEEE Trans. Sustain. Energy **9**(1), 45–55 (2017)
29. Fathy, A.: Butterfly optimization algorithm based methodology for enhancing the shaded photovoltaic array extracted power via reconfiguration process. Energy Convers. Manage. **220**, 113115 (2020)
30. Zhao, Z., et al.: A dynamic particles MPPT method for photovoltaic systems under partial shading conditions. Energy Convers. Manage. **220**, 113070 (2020)

Two-Dimensional Nanomaterials for Solar Cell Technology

Ilyas Bouziani, Ismail Essaoudi, and Abdelmajid Ainane

Abstract This chapter is devoted to the theoretical study of the electronic, optical and photovoltaic properties of semiconductor nanomaterials which arouse a great deal of interest for solar cell technology. The calculations carried out throughout the present study are part of the framework of density functional theory, using ab-initio methods. The electronic and optical properties of siligraphene monolayer in pentagonal phase are investigated by employing the augmented plane wave method. It is shown that SiC_2 monolayer has a semiconductor behavior with moderate gap energy. Furthermore, It is found that the studied system exhibits high absorption peaks and low reflectivity in the visible region. Additionally, using Henry's graphical method, it is found that the maximum photovoltaic efficiency of two-dimensional siligraphene is comparable to that of monocrystalline silicon widely used in solar cell technology. These results indicate that two-dimensional nanomaterials are strong candidate for the future generation of photovoltaic solar cells.

Keywords Solar cells · Photovoltaics · Nanomaterials · Siligraphene · Density functional theory · Augmented plane wave

I. Bouziani (✉) · I. Essaoudi · A. Ainane
Laboratoire de Physique des Matériaux et Modélisation des Systèmes, (LP2MS) Unité Associée au CNRST-URAC 08, Faculty of Sciences, Physics Department, Moulay Ismail University of Meknes, B.P. 11201 Meknes, Morocco
e-mail: i.bouziani@edu.umi.ac.ma

I. Essaoudi
e-mail: i.essaoudi@fs-umi.ac.ma

A. Ainane
e-mail: a.ainane@fs-umi.ac.ma

A. Ainane
Max-Planck-Institut Für Physik Complexer Systeme, Nöthnitzer Str. 38, 01187 Dresden, Germany

S. El Himer et al. (eds.), *Artificial Intelligence of Things for Smart Green Energy Management*, Studies in Systems, Decision and Control 446,
https://doi.org/10.1007/978-3-031-04851-7_7

1 Introduction

Nowadays, efficiently exploiting the green and renewable energies in the face of the dominance of fossil fuels is given priority by researchers and international community [1]. There are two driving forces behind this historic transformation. The first is the need to avoid catastrophic climate change. As we all know, one of the biggest problems we face is global warming, which refers to the remarkable average temperature increase on earth and its effects, such as the rise in the level of the sea, the expansion of deserts and extreme weather events. The increase in greenhouse gases, in particular the increase in CO_2 emissions, is considered to be the principal cause of global warming and the use of fossil fuels is the principal culprit of CO_2 emissions (three quarters) [2]. Thus, replacing fossil fuels with green energies is an essential means of reducing CO_2 emissions and avoiding global warming. The second driving force is the reduction in the amount of non-renewable energy sources such as oil, gas and coal, in contrast to the increasing demand for energy due to the accelerated growth of the world population, the improvement of living standard and the development of modern technologies [3]. Therefore, the use of renewable energy sources is inevitable for the future development of modern society.

Green and renewable energies have a small difference in definition, where "green" refers to being environmentally friendly and non-polluting with lower carbon emissions, while "renewable" emphasizes the possibility of be naturally replenished. However, these two terms are often used interchangeably, describing non-polluting and non-fossil energy sources. This type of energy sources includes solar, wind, geothermal, hydroelectric, biomass, tidal and wave powers [4]. Among these energy sources, solar energy has attracted great interest due to its various advantages, for example, It is a gift from the sun, renewable, reliable and noise-free, which can be used anywhere with low maintenance cost. The most popular way to use solar energy is photovoltaic technology, in which energy is directly converted from sunlight into electricity [5]. The devices related to this process of energy conversion are called solar cells, whose production has doubled every two years since 2002, thus becoming the fastest growing energy technology in the world.

During the last decade, rigorous studies have been carried out on the methods of manufacturing solar cells based on crystalline silicon to make them more efficient and economically viable [6]. However, modern technology involves the design of a highly efficient, ultralight and ultrathin solar cell to meet the demand of macroelectronics and microelectronics. Today, perovskite, organic and nanomaterial solar cells have proven to be extremely efficient for photovoltaic devices due to their low production cost and high power conversion efficiency (PCE) [7–15]. Two-dimensional nanomaterials have attracted great attention in nanoscience and nanotechnology fields since the isolation of graphene by mechanical exfoliation [16–18], thanks to their remarkable electronic, mechanical and optical properties leading to many new applications in electronic, optoelectronic, photovoltaic, photocatalytic devices, etc. [19–26]. Since

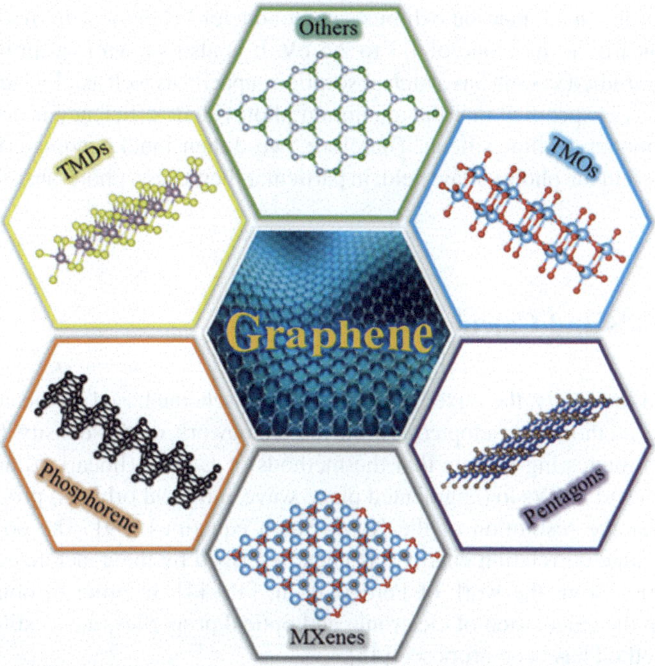

Fig. 1 Schematic illustration of different types of typical two-dimensional nanomaterials

then, a large number of graphene-like structures with various crystal and chemical compositions have been successfully proposed and synthesized, such as transition metal dichalcogenides (TMDs), phosphorene, transition metal carbides/nitrides (MXenes), pentagons, transition metal oxides (TMOs) and others (Fig. 1) [27–33].

In this regard, Zhang et al. [33], have recently proposed a new two-dimensional nanomaterial based on carbon atoms in pentagonal phase, ie pentagraphene, with the hybridization sp^2-sp^3 orbitals, which is mechanically thermally and dynamically stable. Two dimensional pentagraphene possesses an unusual negative Poisson's ratio, ultra-high ideal strength surpassing graphene, and other attractive electronic properties that make it a potential candidate for wide range of applications in electronics and optoelectronics [33, 34]. The silicon counterpart of the pentagonal structure has also been proposed, although it is dynamically unstable in two-dimensional form [35]. More recently, a new pentagonal structures composed of sp^2 hybridized carbon atoms and sp^3 hybridized silicon atoms, i.e.: siligraphene (SiC_2), was confirmed as being energetically and dynamically stable, and exhibiting improved electronic properties compared to those of pentagraphene, in particular the semiconductor behavior necessary for photovoltaic applications [36–38].

The main objective of this chapter is to study theoretically the electronic, optical and photovoltaic properties of two-dimensional pentagonal-type nanomaterials, focusing on siligraphene compound, for photovoltaic technology. The finding results indicate that this nanomaterial exhibits semiconductor behavior with moderate gap energy, typically in the range of 1.3 to 2.5 eV. It is also shown that, in the visible domain, the studied system has a high absorption capacity as well as a low reflectivity. In addition, it is specified that the maximum photovoltaic efficiency is comparable to that of monocrystalline silicon. Therefore, two-dimensional nanomaterials are of major interest in the photovoltaic field, in particular their use as photo-absorber layers in solar cells.

2 Methods and Calculation Details

To carry out this study, the mixed LAPW/APW+lo scheme based on the augmented plane wave method was adopted within the framework of the density functional theory. It is interesting to note that the methods of LAPW (linearized augmented plane wave) and APW+lo (augmented plane wave and local orbitals) propose basis functions for the resolution of the Kohn–Sham equations [39]. The contribution of the exchange correlation energy has been described by the generalized gradient approximation from the work of Perdew et al. [40–44]. In order to obtain better precision in the calculation of electronic and optical properties, the modified Beck-Jonhson method has been proposed [45].

It should be remembered that the first step in this type of computation consists in specifying the parameters value which influence the precision of results [46], namely:

- Muffin-tin radius (R_{MT}). The choice the R_{MT} values for Si and C atoms is based on two main criteria:
  - Ensure the integration of the core electrons majority in the muffin-tin spheres.
  - Avoid overlapping muffin-tin spheres.
- The cutoff parameter ($R_{MT}K_{\max}$), determining the number of basis functions, with R_{MT} is the smallest radius of the muffin-tin spheres in the unit cell and $K_{\max}$ is the magnitude of the largest wave vector utilized for the plane wave expansion of the eigenfunctions.
- The parameter $l_{\max}$ which represents the maximum angular momentum of the basis functions for the atomic orbitals (inside the sphere).
- The parameter $G_{\max}$ which indicates the (magnitude of largest wave vector utilized in the Fourier expansion of the charge density $\rho(r) = \sum_{G}^{G_{\max}} A_G e^{iG.r}$.
- The k–points number representing the mesh of the 1st Brillouin zone.

We point out that the number of k–points is not a criterion of convergence towards minimum energy, but rather a criterion of convergence stability. Indeed the behavior of the total energy according to the k–points is not regular (the number of the k–points is not a variational parameter like the density) and therefore it gives total energies which go erratically towards the exact value [47].

It should also be noted that the R_{MT} radii for different atoms should not be so different; this will make it difficult to find an adequate l_{max} for each atom in the cell. An acceptable margin of difference would be 10 to 20% for d, p and s states [46].

The plane wave functions inside a muffin-tin sphere are developed in terms of spherical waves (which gives them the name of augmented plane waves), with a cutoff angular momentum of the spherical harmonics l_{max}. In the interstitial region, plane waves describe atomic valence orbitals up to a cutoff wave vector K_{max}. The wave functions are continuous at the boundary of the muffin-tin sphere, and no approximation has been imposed on the shape of the crystal charge density or potential.

In the calculations carried out in this chapter, $R_{MT} K_{max} = 7$, $l_{max} = 10$ and $G_{max} = 14 Ry^{1/2}$ were adopted for the studied structure, in order to compete with the precision reported in the literature. The space reciprocal was also sampled using up to 500 K-points in the irreducible part of the 1st Brillouin zone to perform structural optimization and to determine electronic properties, while up to 700 K-points were used in order to find the optical ones of the considered compound. Furthermore, the error in the total energy between two successive self-coherent cycles has been fixed at $10^{-5} Ry$ in order to ensure convergence.

The determination of the geometry of structures at equilibrium is carried out by minimizing the total energy as a function of the unit cell volume. The total energy values obtained for various volumes were interpolated utilizing the Murnaghan equation [48], given by:

$$B(V) = E_0 + \frac{B_0 V}{B_0^{'}} \left[\frac{(V_0/V)^{B_0^{'}}}{B_0^{'} - 1} + 1 \right] - \frac{B_0 V_0}{B_0^{'} - 1}, \tag{1}$$

where E_0, B_0 and V_0 are the total energy, the compressive modulus and the volume at equilibrium of the unit cell, respectively. $B_0^{'}$ is the derivative of the compression modulus with respect to the equilibrium pressure.

It is also important to note that the difference between the atomic layers is fixed at 15 Å in the "z" direction for the system considered, in order to avoid the effects of structures in periodic layers [48, 49].

3 Results and Discussion

3.1 Electronic Properties

We begin our investigation by presenting the results of the crystal structure optimization of two-dimensional siligraphene compound in tetragonal phase, which has the P-421m symmetry (space group n° 113). To do this, the lattice parameters *a* and *b*, as well as the atomic coordinates were relaxed by performing self-consistent computation of the total energy for the system under consideration within the framework of the generalized gradient approximation.

In Fig. 2, we present the variations of the total energy as a function of the volume for elementary cell of SiC_2 system, using the adjustment by the Murnaghan equation of state (Eq. 1). It is found that the optimized lattice parameters (a; b) are (4.405 Å; 4.405 Å), as illustrated in Table 1. The values of the lattice parameters a and b as well as the thickness d are in good agreement with those obtained theoretically in the literature [50]. Thus, the optimized structure of two-dimensional SiC_2 system is shown in Fig. 3.

After determining the optimal structure of two-dimensional siligraphene system, its electronic properties were investigated based on the calculation of the projected and total electronic density of states, using the generalized gradient approximation corrected by the modified Beck-Jonhson potential.

Figures 4 and 5, illustrate, respectively, the projected and total electronic density of states of two-dimensional SiC_2 compound in the energy range of −3 to 3 eV. It is found that SiC_2 monolayer shows semiconductor behavior with moderate gap energy

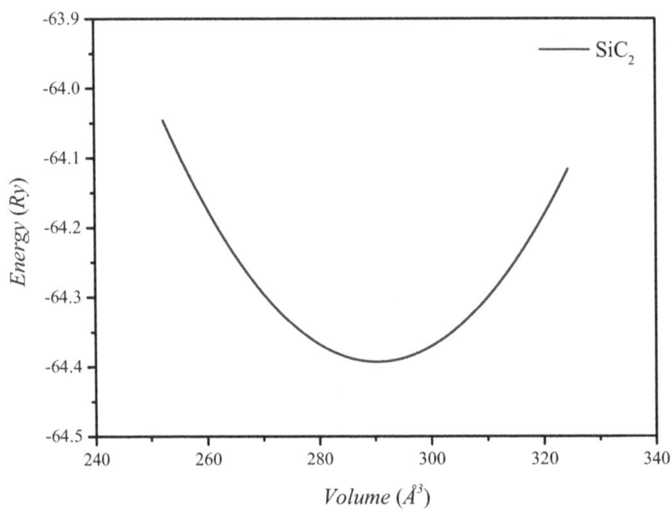

Fig. 2 Variation of total energy as a function of unit cell volume for two-dimensional siligraphene system

Table 1 Lattice (a; b), thickness (d) and gap energy (E_g) parameters of siligraphene system monolayer

Compound	a(Å)	b(Å)	d(Å)	E_g (eV)
SiC_2	4.405 4.412[x]	4.405 4.412[x]	1.329 1.320[x]	1.75 1.429[x]

[x] Ref. [50]

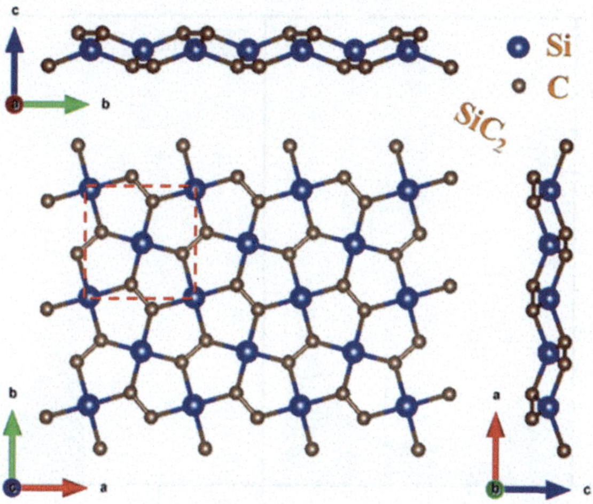

Fig. 3 Top and side views of the optimized siligraphene system. The red dashed area represents the unit cell

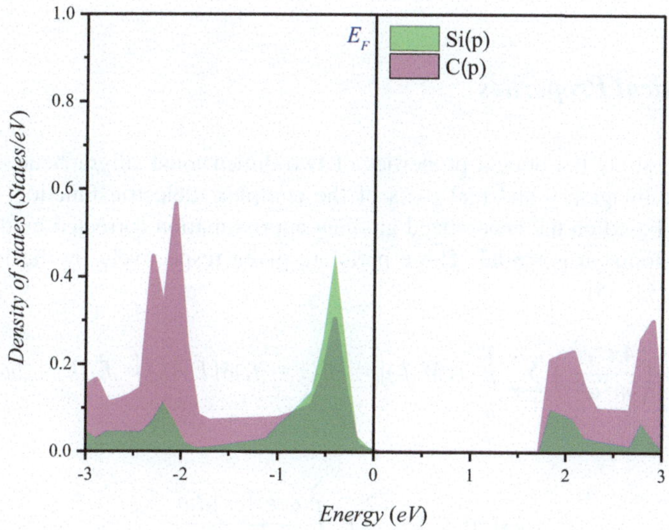

Fig. 4 Projected density of states of siligraphene system. The Fermi level is taken as the origin of the energies

of 1.75 eV (see Table 1). These results are in good agreement with the literature where SiC_2 exhibits semiconductor behavior referring to the Heyd, Scuseria, Ernzerhof (HSE) approximation [50, 51]. Also, It is clear that the valence band is principally formed by the 2p(C) and 3p(Si) states, while the conduction band is mainly dominated by the 2p(C) state.

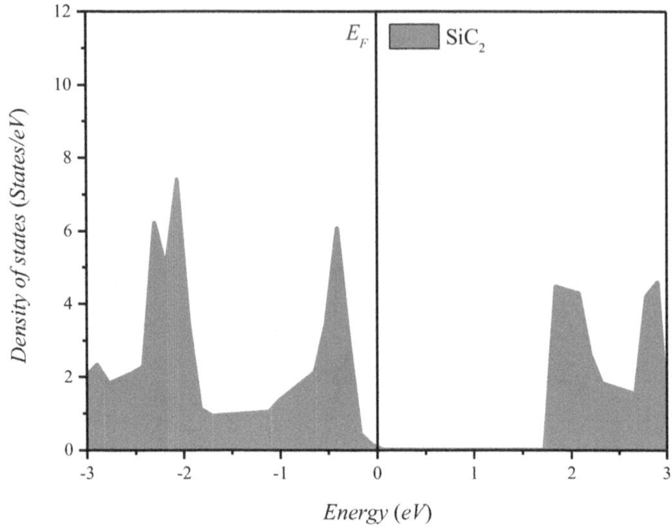

Fig. 5 Total density of state of siligraphene system. The Fermi level is taken as the origin of the energies

3.2 Optical Properties

In order to study the optical properties of two-dimensional siligraphene semiconductor, the imaginary and real parts of the complex dielectric function (ε) were calculated based on the generalized gradient approximation corrected by the modified Beck-Jonhson potential. These parts are given respectively by the following equations [52–55]:

$$\varepsilon_2(\omega) = \left(\frac{4\pi^2 e^2}{m^2\omega^2}\right)\sum_{i,j}\int_{ZB} \langle i|\,M(k)|\,j\rangle^2 f_i(1-f_i)\delta\big(E_f(k) - E_i(k) - \hbar\omega\big)d^3k, \tag{2}$$

$$\varepsilon_1(\omega) = 1 + \frac{2}{\pi}P\int_0^\infty \frac{\omega'\varepsilon_2(\omega')d\omega'}{\omega'^2 - \omega^2}, \tag{3}$$

where the parameters ω, M and k are respectively the frequency of the incident photons, the dipole matrix and the reciprocal lattice vector. The indices i and j denote the unoccupied and occupied states, respectively. f_i and E_i are, respectively, the Fermi distribution and the energy of the electron in the ith state. Also, P denotes the principal Cauchy value of the integral defined by:

$$P = \lim_{a \to 0} \int_{-\infty}^{\omega - a} \frac{\omega' \varepsilon(\omega') d\omega'}{\omega' - \omega} + \lim_{a \to 0} \int_{\omega + a}^{+\infty} \frac{\omega' \varepsilon(\omega') d\omega'}{\omega' - \omega}. \tag{4}$$

In Fig. 6, the variations of the real and imaginary parts of dielectric function versus the photon energy have been shown for the present system. For the variations of the real part $\varepsilon_1(\omega)$, the most important quantity is the dielectric constant $\varepsilon_1(0)$ which strongly depends on the gap energy [56]. This quantity is equal to 3.07 for SiC_2 monolayer. For the variation of the imaginary part $\varepsilon_2(\omega)$, the SiC_2 shows two main peaks at the energies $E_1 = 2.34$ eV and $E_2 = 3.16$ eV in the visible region, which correspond to the transitions between the 2p(C) → 2p(C) and 3p(Si) → 2p(C) orbitals, due to the dominance of these orbitals at the valence and conduction bands according to the projected and total electronic density of states for this compound.

For the purpose of giving more details about the optical properties of siligraphene monolayer, the absorption coefficients $\alpha(\omega)$ and reflectivity $R(\omega)$ for the examined system were respectively calculated using the following equations [55, 57, 58]:

$$\alpha(\omega) = \sqrt{2}\omega \left[\sqrt{\varepsilon_1^2(\omega) + \varepsilon_2^2(\omega)} + \varepsilon_1(\omega) \right]^{1/2} \tag{5}$$

$$R(\omega) = \left| \frac{\sqrt{\varepsilon_1(\omega) + i\varepsilon_2(\omega)} - 1}{\sqrt{\varepsilon_1(\omega) + i\varepsilon_2(\omega)} + 1} \right|^2 \tag{6}$$

In Fig. 7, the absorption spectrum of the investigated compound and the solar

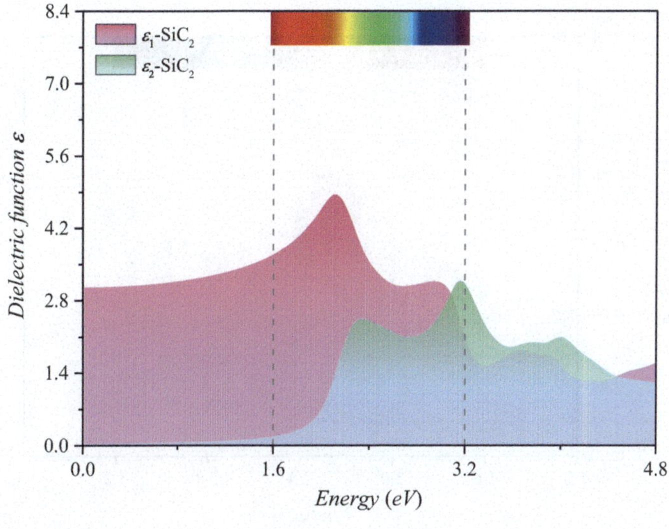

Fig. 6 The real and imaginary parts of dielectric function of two-dimensional siligraphene system versus the photon energy

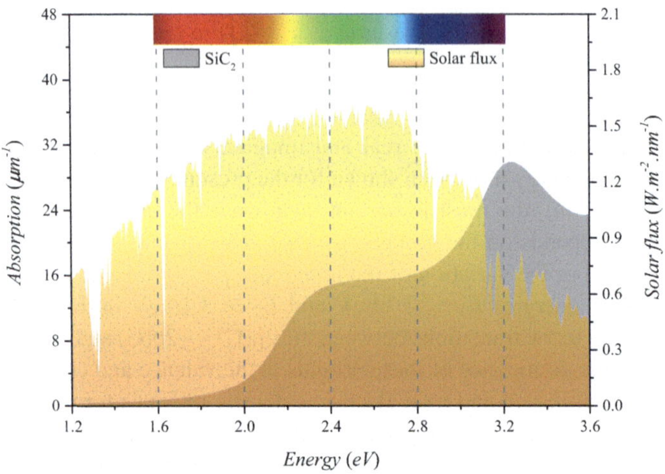

Fig. 7 Variation of the optical absorption spectrum of siligraphene system versus the photon energy. AM1.5 solar spectrum is also provided [61]

flux spectrum between 1.2 and 3.6 eV have been illustrated. We notice that in the visible region, SiC_2 compound exhibits two high absorption peaks reaching up to 30 μm^{-1}). This maximum absorption value are higher than that obtained for hybrid perovskite $CH_3NH_3PbI_3$ (~26 μm^{-1}) recently proposed as absorber semiconductor for photovoltaic applications [59].

In addition, Fig. 8 shows the reflectivity spectrum of the considered system. It

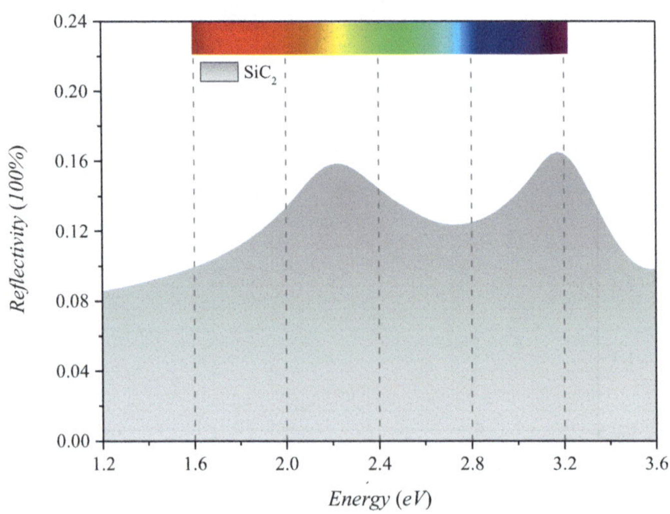

Fig. 8 Variation of the reflectivity spectrum of two-dimensional siligraphene system versus photon energy

should be noted that the reflectivity of SiC_2 monolayer does not exceed 18% in the visible range, so that the highest reflectivity value is observed around 2.08 eV. The maximum value of reflexivity is comparable to that of Cs_2PtI_6 and Rb_2PtI_6 which are highly desirable, as double perovskite compounds, for photovoltaic applications [60].

3.3 *Photovoltaic Efficiency of Two-Dimensional Semiconductors*

Let us now move on to the estimation of the photovoltaic efficiencies of the siligraphene monolayer studied in this chapter. To do it, the graphic method approached in Henry works is used [62]. For this method, the photovoltaic efficiency of a semiconductor is expressed by:

$$E_{eff} = \frac{100 n_{ph} W}{P/S}\% = \frac{100 n_{ph} W}{N_{ph}\langle\hbar\omega\rangle}\%. \tag{7}$$

where the terms n_{ph}, W and P/S are the solar flux absorbed by the semiconductor, the work done by the absorbed photons and the total solar power per unit area, respectively. Moreover, the term $\langle\hbar\omega\rangle$ represents the average energy of the photons and N_{ph} equal to n_{ph} for $E_g = 0eV$.

For the AM1.5[1] solar spectrum received at the earth surface at mid-latitudes, the function $n_{ph}(E_g)$ is related to the gap energy E_g of the semiconductor by the formula:

$$n_{ph}(E_g) = \int_{E_g}^{\infty} \frac{dn_{ph}}{d\hbar\omega} d\hbar\omega. \tag{8}$$

Also, the work W is expressed as follows:

$$W = \frac{JV}{n_{ph}}, \tag{9}$$

where J and V are successfully the current density delivered by an ideal semiconductor-based solar cell and the voltage at the bounds of the load, with:

$$J = en_{ph} - A\exp\left(\frac{eV - E_g}{kT}\right). \tag{10}$$

[1] The solar spectrum AM1.5, for Air Mass, means that atmospheric conditions are considered to be clear and that sunlight passes through 1.5 times the actual thickness of the atmosphere due to inclination, which corresponds to a latitude of 48° at noon.

Thus, the point of maximum power ($J_{\max}, V_{\max}$) is obtained by taking the derivative $dJ/dV = 0$, which leads to the following formulas:

$$eV_{\max} = eV_{oc} - kTln\left(1 + \frac{eV_{\max}}{kT}\right), \tag{11}$$

$$J_{\max} = \frac{en_{ph}}{1 + kT/eV_{\max}}, \tag{12}$$

where V_{oc} is the open circuit voltage, obtained for J equal to zero:

$$eV_{oc} = E_g - kTln\left(\frac{A}{en_{ph}}\right). \tag{13}$$

With, $A = 5693E_g^2$ (A/cm^2) and $kT = 0.0258\,\text{eV}$.

Therefore, the maximum work done by the absorbed photons is given by:

$$W_{\max} = \frac{J_{\max}V_{\max}}{n_{ph}} \cong eV_{\max} - kT. \tag{14}$$

This expression is used in Eq. (7) to calculate the maximum photovoltaic efficiency of two-dimensional semiconductors, after having graphically determined the values of n_{ph} for each gap energy from Fig. 9.

Table 2 illustrates the results of the maximum photovoltaic efficiency E_{eff} of two-dimensional SiC_2 semiconductor, as well as the absorbed solar flux n_{ph} and

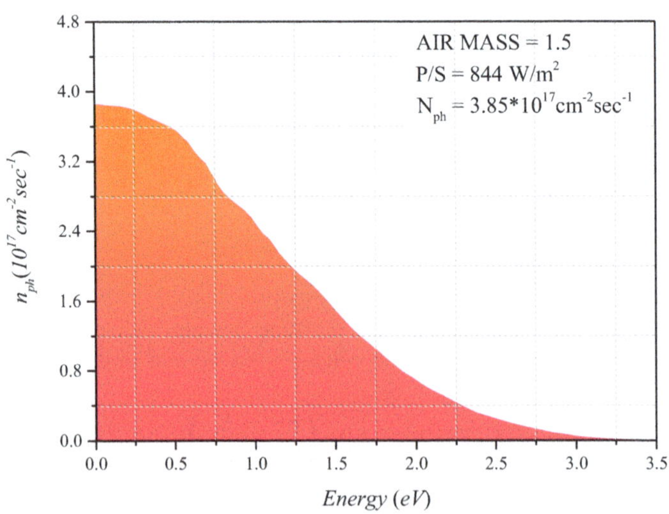

Fig. 9 Photons flux absorbed by the ideal solar cell versus gap energy [62]

Table 2 Absorbed solar flux (n_{ph}), maximum work done by absorbed photons (W_{max}) and maximum photovoltaic efficiency (E_{eff}) of two-dimensional siligraphene semiconductor

Compound	$n_{ph}(cm^{-2}s^{-1})$	W_{max} (eV)	E_{eff} (%)
SiC_2	1.06×10^{17}	1.27	25.56

the maximum work performed by the absorbed photons W_{max} for the examined system. It is mainly observed that the maximum value of the photovoltaic efficiency of siligraphene semiconductor equal to 25.56%. This value is relatively comparable to that of crystalline silicon of 26.06% [63–65].

4 Conclusion

In this chapter we have presented the electronic, optical and photovoltaic properties of two-dimensional siligraphene of pentagonal type using ab-initio calculations based on the augmented plane wave method, with the mixed scheme LAPW/APW +lo, in the framework of the generalized gradient approximation corrected by the modified Beck-Jonhson potential.

For the study of the electronic properties, the densities of states (projected and total) were determined for SiC_2 compound. It is shown that this system has a semiconductor behavior with moderate gap energy of 1.75 eV. It is further seen that the valence band is principally formed by the 2p(C) and 3p (Si) states, while the conduction band is mainly dominated by the 2p(C) state.

Concerning the optical properties, the absorption and reflectivity parameters have been principally computed for the considered system. It is essentially found that this nanomaterial exhibits high absorption peaks up to 30 μm^{-1} in the visible region. It is also observed that the maximum reflectivity does not exceed 18% in the visible range.

Moreover, by appealing to the graphical method of Henry and by using the gap energy obtained for the examined system, it is found that the maximum value of the photovoltaic efficiency of SiC_2 semiconductor equal to 25.56%, which is comparable to that of monocrystalline silicon widely used in solar cell technology.

Thus, the two-dimensional siligraphene is an excellent candidate for the future generation of photovoltaic solar cells.

Acknowledgements The present work has been initiated with the support of the project PPR2: (MESRSFC-CNRST).

References

1. Goswami, D.Y., Zhao, Y.: Proceedings of ISES World Congress 2007: Solar Energy and Human Settlement, vol. 1. Springer Science & Business Media (2009)
2. Bailey, M., Hallett, J.: Nucleation effects on the habit of vapour grown ice crystals from −18 to −42 C. Q. J. R. Meteorol. Soc. J. Atmos. Sci. Appl. Meteorol. Phys. Oceanogr. **128**, 1461–1483 (2002)
3. Petroleum, B.: British petroleum statistical review of world energy. British Petroleum Corporate Communications Services, London (2001)
4. Ellabban, O., Abu-Rub, H., Blaabjerg, F.: Renewable energy resources: current status, future prospects and their enabling technology. Renew. Sustain. Energy Rev. **39**, 748–764 (2014)
5. Becquerel, P.: Comptes Rendus de l'Academie des Sciences. Paris **9**, 561 (1839)
6. Saga, T.: Advances in crystalline silicon solar cell technology for industrial mass production. NPG Asia Mater. **2**, 96–102 (2010)
7. Carey, G.H., Abdelhady, A.L., Ning, Z., Thon, S.M., Bakr, O.M., Sargent, E.H.: Colloidal quantum dot solar cells. Chem. Rev. **115**, 12732–12763 (2015)
8. Green, M.A., Ho-Baillie, A., Snaith, H.J.: The emergence of perovskite solar cells. Nat. Photonics **8**, 506–514 (2014)
9. Krebs, F.C., Gevorgyan, S.A., Alstrup, J.: A roll-to-roll process to flexible polymer solar cells: model studies, manufacture and operational stability studies. J. Mater. Chem. **19**, 5442–5451 (2009)
10. Linghu, J., Yang, T., Luo, Y., Yang, M., Zhou, J., Shen, L., Feng, Y.P.: High-throughput computational screening of vertical 2D van der Waals heterostructures for high-efficiency excitonic solar cells. ACS Appl. Mater. Interfaces **10**, 32142–32150 (2018)
11. Scharber, M.C., Sariciftci, N.S.: Efficiency of bulk-heterojunction organic solar cells. Prog. Polym. Sci. **38**, 1929–1940 (2013)
12. Snaith, H.J.: Perovskites: the emergence of a new era for low-cost, high-efficiency solar cells. J. Phys. Chem. Lett. **4**, 3623–3630 (2013)
13. Tsai, M.-L., Su, S.-H., Chang, J.-K., Tsai, D.-S., Chen, C.-H., Wu, C.-I., Li, L.-J., Chen, L.-J., He, J.-H.: Monolayer MoS_2 heterojunction solar cells. ACS Nano **8**, 8317–8322 (2014)
14. Tsakalakos, L., Balch, J., Fronheiser, J., Korevaar, B.A., Sulima, O., Rand, J.: Silicon nanowire solar cells. Appl. Phys. Lett. **91**, 233117 (2007)
15. Yang, B., Wang, M., Hu, X., Zhou, T., Zang, Z.: Highly efficient semitransparent CsPbIBr2 perovskite solar cells via low-temperature processed In2S3 as electron-transport-layer. Nano Energy **57**, 718–727 (2019)
16. Novoselov, K.S., Geim, A.K., Morozov, S.V., Jiang, D., Zhang, Y., Dubonos, S.V., Grigorieva, I.V., Firsov, A.A.: Electric field effect in atomically thin carbon films. Science **306**, 666–669 (2004)
17. Novoselov, K.S., Geim, A.K., Morozov, S.V., Jiang, D., Katsnelson, M.I., Grigorieva, Iv., Dubonos, Sv., Firsov, A.A.: Two-dimensional gas of massless Dirac fermions in graphene. Nature **438**, 197–200 (2005)
18. Zhang, Y., Tan, Y.-W., Stormer, H.L., Kim, P.: Experimental observation of the quantum Hall effect and Berry's phase in graphene. Nature **438**, 201–204 (2005)
19. Balandin, A.A.: Thermal properties of graphene and nanostructured carbon materials. Nat. Mater. **10**, 569–581 (2011)
20. Balandin, A.A., Ghosh, S., Bao, W., Calizo, I., Teweldebrhan, D., Miao, F., Lau, C.N.: Superior thermal conductivity of single-layer graphene. Nano Lett. **8**, 902–907 (2008)
21. Schedin, F., Geim, A.K., Morozov, S.V., Hill, E.W., Blake, P., Katsnelson, M.I., Novoselov, K.S.: Detection of individual gas molecules adsorbed on graphene. Nat. Mater. **6**, 652–655 (2007)

22. Benhouria, Y., Bouziani, I., Essaoudi, I., Ainane, A., Ahuja, R.: Quantum Monte Carlo study of dynamic magnetic properties of nano-graphene. J. Magn. Magn. Mater. **460**, 223–228 (2018)
23. Haman, Z., Kibbou, M., Bouziani, I., Benhouria, Y., Essaoudi, I., Ainane, A., Ahuja, R.: Structural, electronic and optical properties of two-dimensional Janus transition metal oxides MXO (M= Ti, Hf and Zr; X= S and Se) for photovoltaic and opto-electronic applications. Physica B **604**, 412621 (2021)
24. Kibbou, M., Haman, Z., Bouziani, I., Khossossi, N., Benhouria, Y., Essaoudi, I., Ainane, A., Ahuja, R.: Cs2InGaX6 (X= Cl, Br, or I): emergent inorganic halide double perovskites with enhanced optoelectronic characteristics. Curr. Appl. Phys. **21**, 50–57 (2021)
25. Bouziani, I., Kibbou, M., Haman, Z., Benhouria, Y., Essaoudi, I., Ainane, A., Ahuja, R.: Electronic and optical properties of ZnO nanosheet doped and codoped with Be and/or Mg for ultraviolet optoelectronic technologies: density functional calculations. Phys. Scr. **95**, 015804 (2020)
26. Bouziani, I., Kibbou, M., Haman, Z., Khossossi, N., Essaoudi, I., Ainane, A., Ahuja, R.: Two-dimensional Janus Sn_2SSe and $SnGeS_2$ semiconductors as strong absorber candidates for photovoltaic solar cells: first principles computations. Physica E **134**, 114900 (2021)
27. Castellanos-Gomez, A., Barkelid, M., Goossens, A.M., Calado, V.E., van der Zant, H.S., Steele, G.A.: Laser-thinning of MoS_2: on demand generation of a single-layer semiconductor. Nano Lett. **12**, 3187–3192 (2012)
28. Haque, F., Daeneke, T., Kalantar-Zadeh, K., Ou, J.Z.: Two-dimensional transition metal oxide and chalcogenide-based photocatalysts. Nano-micro Lett. **10**, 1–27 (2018)
29. Qiao, J., Kong, X., Hu, Z.-X., Yang, F., Ji, W.: High-mobility transport anisotropy and linear dichroism in few-layer black phosphorus. Nat. Commun. **5**, 1–7 (2014)
30. Xia, F., Wang, H., Jia, Y.: Rediscovering black phosphorus as an anisotropic layered material for optoelectronics and electronics. Nat. Commun. **5**, 1–6 (2014)
31. Zha, X.-H., Huang, Q., He, J., He, H., Zhai, J., Francisco, J.S., Du, S.: The thermal and electrical properties of the promising semiconductor MXene Hf_2CO_2. Sci. Rep. **6**, 1–10 (2016)
32. Zhang, W., Huang, Z., Zhang, W., Li, Y.: Two-dimensional semiconductors with possible high room temperature mobility. Nano Res. **7**, 1731–1737 (2014)
33. Zhang, X., Zhao, X., Wu, D., Jing, Y., Zhou, Z.: High and anisotropic carrier mobility in experimentally possible Ti_2CO_2 (MXene) monolayers and nanoribbons. Nanoscale **7**, 16020–16025 (2015)
34. Xu, W., Zhang, G., Li, B.: Thermal conductivity of penta-graphene from molecular dynamics study. J. Chem. Phys. **143**, 154703 (2015)
35. Aierken, Y., Leenaerts, O., Peeters, F.M.: A first-principles study of stable few-layer penta-silicene. Phys. Chem. Chem. Phys. **18**, 18486–18492 (2016)
36. Bouziani, I., Haman, Z., Kibbou, M., Benhouria, Y., Essaoudi, I., Ainane, A., Ahuja, R.: Ab initio study of electronic and optical properties of penta-SiC_2 and-$SiGeC_4$ monolayers for solar energy conversion. Superlattices Microstruct. **142**, 106524 (2020)
37. Bouziani, I., Haman, Z., Kibbou, M., Essaoudi, I., Ainane, A., Ahuja, R.: Electronic, optical and thermoelectric properties of two-dimensional pentagonal $SiGeC_4$ nanosheet for photovoltaic applications: first-principles calculations. Superlattices Microstruct. **158**, 107024 (2021)
38. Lopez-Bezanilla, A., Littlewood, P.B.: σ–π-band inversion in a novel two-dimensional material. J. Phys. Chem. C **119**, 19469–19474 (2015)
39. Blaha, P., Schwarz, K., Sorantin, P., Trickey, S.B.: Full-potential, linearized augmented plane wave programs for crystalline systems. Comput. Phys. Commun. **59**, 399–415 (1990)
40. Perdew, J.P., Burke, K., Ernzerhof, M.: Generalized gradient approximation made simple. Phys. Rev. Lett. **77**, 3865 (1996)
41. Bouziani, I., Benhouria, Y., Essaoudi, I., Ainane, A., Ahuja, R.: Magnetoelectronic properties of GaN codoped with (V, Mn) impurities for spintronic devices: ab-initio and Monte Carlo studies. Physica A **512**, 1249–1259 (2018)

42. Bouziani, I., Benhouria, Y., Essaoudi, I., Ainane, A., Ahuja, R.: Magnetoelectronic properties of Vanadium impurities co-doped (Cd, Cr) Te compound for spintronic devices: first principles calculations and Monte Carlo simulation. J. Magn. Magn. Mater. **466**, 420–429 (2018)
43. Bouziani, I., Benhouria, Y., Essaoudi, I., Ainane, A., Ahuja, R.: Half metallic ferromagnetic behavior in (Ga, Cr) N and (Ga, Cr, V) N compounds for spintronic technologies: Ab initio and Monte Carlo methods. J. Magn. Magn. Mater. **477**, 220–225 (2019)
44. Bouziani, I., Benhouria, Y., Essaoudi, I., Ainane, A., Ahuja, R.: High Curie temperature in halfmetallic ferromagnets (Zn, Cr, Ti) Se and (Zn, Cr, Ti) Te for spintronic devices: ab initio and Monte Carlo treatments. Mater. Sci. Eng. B **253**, 114484 (2020)
45. Tran, F., Blaha, P.: Accurate band gaps of semiconductors and insulators with a semilocal exchange-correlation potential. Phys. Rev. Lett. **102**, 226401 (2009)
46. Blaha, P., Schwarz, K., Madsen, G.K., Kvasnicka, D., Luitz, J.: Wien2k. An augmented plane wave+ local orbitals program for calculating crystal properties **60** (2001)
47. Cottenier, S.: Density Functional Theory and the family of (L) APW-methods: a step-by-step introduction. Instituut voor Kern-en Stralingsfysica, KU Leuven, Belgium **4**, 41 (2002)
48. Li, Z., Xu, W., Yu, Y., Du, H., Zhen, K., Wang, J., Luo, L., Qiu, H., Yang, X.: Monolayer hexagonal arsenene with tunable electronic structures and magnetic properties via impurity doping. J. Mater. Chem. C **4**, 362–370 (2016)
49. Zheng, Z., Wang, X., Mi, W.: Electric field tunable electronic structure in two dimensional van der Waals g-C2N/XSe2 (X= Mo, W) heterostructures. Carbon **117**, 393–398 (2017)
50. Berdiyorov, G.R., Madjet, M.E.-A.: First-principles study of electronic transport and optical properties of penta-graphene, penta-SiC_2 and penta-CN_2. RSC Adv. **6**, 50867–50873 (2016)
51. Li, X., Dai, Y., Li, M., Wei, W., Huang, B.: Stable Si-based pentagonal monolayers: high carrier mobilities and applications in photocatalytic water splitting. J. Mater. Chem. A **3**, 24055–24063 (2015)
52. Peter, Y.U., Cardona, M.: Fundamentals of Semiconductors: Physics and Materials Properties. Springer Science & Business Media (2010)
53. Saha, S., Sinha, T.P., Mookerjee, A.: Electronic structure, chemical bonding, and optical properties of paraelectric $BaTiO_3$. Phys. Rev. B **62**, 8828 (2000)
54. Read, A.J., Needs, R.J.: Calculation of optical matrix elements with nonlocal pseudopotentials. Phys. Rev. B **44**, 13071 (1991)
55. Smith, D.Y.: Dispersion theory, sum rules, and their application to the analysis of optical data. In: Handbook of Optical Constants of Solids (1997), pp. 35–68
56. Alemi, A., Hosseinpour, Z., Dolatyari, M., Bakhtiari, A.: Boehmite (γ-AlOOH) nanoparticles: hydrothermal synthesis, characterization, pH-controlled morphologies, optical properties, and DFT calculations. Physica Status Solidi (B) **249**, 1264–1270 (2012)
57. Bass, M., Van Stryland, E.W., Williams, D.R., Wolfe, W.L.: Handbook of optics, vol. 2. McGraw-Hill New York (1995)
58. Peyghambarian, N., Koch, S.W., Mysyrowicz, A.: Introduction to semiconductor optics. Prentice-Hall, Inc. (1993)
59. Shirayama, M., Kadowaki, H., Miyadera, T., Sugita, T., Tamakoshi, M., Kato, M., Fujiseki, T., Murata, D., Hara, S., Murakami, T.N.: Optical transitions in hybrid perovskite solar cells: ellipsometry, density functional theory, and quantum efficiency analyses for $CH_3NH_3PbI_3$. Phys. Rev. Appl. **5**, 014012 (2016)
60. Zhao, X.-H., Wei, X.-N., Tang, T.-Y., Gao, L.-K., Xie, Q., Lu, L.-M., Tang, Y.-L.: First-principles study on the structural, electronic and optical properties of vacancy-ordered double perovskites Cs2PtI6 and Rb2PtI6. Opt. Mater. **114**, 110952 (2021)
61. Kurucz, R.L.: Solar Flux Atlas from 296 to 1300 nm. National Solar Observatory Atlas, 1 (1984)
62. Henry, C.H.: Limiting efficiencies of ideal single and multiple energy gap terrestrial solar cells. J. Appl. Phys. **51**, 4494–4500 (1980)

63. Kayes, B.M., Nie, H., Twist, R., Spruytte, S.G., Reinhardt, F., Kizilyalli, I.C., Higashi, G.S.: 27.6% Conversion efficiency, a new record for single-junction solar cells under 1 sun illumination. In: 2011, 37th IEEE Photovoltaic Specialists Conference, (2011), 000004–000008
64. Green, M., Dunlop, E., Hohl-Ebinger, J., Yoshita, M., Kopidakis, N., Hao, X.: Solar cell efficiency tables (version 57). Prog. Photovoltaics Res. Appl. **29**, 3–15 (2021)
65. Green, M.A., Hishikawa, Y., Warta, W., Dunlop, E.D., Levi, D.H., Hohl-Ebinger, J., Ho-Baillie, A.W.H.: Solar cell efficiency tables (version 50). Prog. Photovoltaics Res. Appl. **25**, 668–676 (2017)

Improving Control for an Induction Machine Using Artificial Intelligence

Zineb Mekrini, Sara Mensou, Mhamed El Mrabet, and Mohammed Boulaala

Abstract The performance of the induction machine control system has not always been guaranteed due to the complexity of the nonlinear characteristics of these machines. For this, we adopted a model of the control adapted to the identifiable parameters. The control method developed is the Direct Torque Control or DTC. This control is based on simple hysteresis regulators and switching tables, providing energy advantages in both regimes, but presents ripples in torque and flux. In order to minimize the ripples, we have developed a technique for improving the classical structure of DTC based on Artificial Intelligence using Fuzzy Logic.

Keywords Induction machine · Artificial intelligence · DTC control · Fuzzy controller

1 Introduction

The induction machine enjoys particular attention in modern industrial installations for several reasons, in particular: its low cost, its reduced mass, its robustness and its simplicity of manufacture and maintenance [1]. However, this simplicity is however accompanied by a great complexity of control, because of the coupling between the stator and rotor fields, requiring much more complex control techniques than the DC

Z. Mekrini (✉) · M. El Mrabet · M. Boulaala
Laboratory of Informatics, Systems & Telecommunications (LIST), Faculty of Science and Techniques of Tangier (FST), Abdelmalek-Essaadi University (UAE), Tetouan, Morocco
e-mail: zmekrini@uae.ac.ma

M. El Mrabet
e-mail: melmrabet@uae.ac.ma

M. Boulaala
e-mail: m.boulaala@uae.ac.ma

S. Mensou
Research Centre of Engineering and Health Sciences and Technologies (STIS), ENSAM Mohammed V University, Rabat, Morocco
e-mail: sara.mensou11@gmail.com

S. El Himer et al. (eds.), *Artificial Intelligence of Things for Smart Green Energy Management*, Studies in Systems, Decision and Control 446,
https://doi.org/10.1007/978-3-031-04851-7_8

machine [2]. Also, that the induction machine requires stator currents all the time even at zero torque to maintain the rotor flux.

To remedy this limitation of induction motors, it was necessary to think of a control strategy, which would ensure decoupling of these motors and adjustment flexibility similar to that of direct current drives. The Direct Control techniques were originally based on a simplified knowledge of machine behavior. Tuning methods are hysteresis comparators and selection tables [3]. This control strategy allows to calculate the control quantities which are the electromagnetic torque and the stator flux from the only quantities related to the stator with or without the intervention of a mechanical sensor [4]. The hysteresis comparator leads to operation at variable switching frequency leading to risks of mechanical excitation [5]. This variation causing an error in the estimation of the torque and the stator flux, which can cause the instability of the DTC [6]. For this, we will develop several techniques, in order to improve performance of DTC in this chapter which has four sections:

In the first section the electrical and mechanical construction of an Induction Machine is detailed. We present the mathematical model of the induction machine in the two-phase frame linked to the stator. The results of the simulation of this machine supplied with voltage by a three-phase are illustrated.

The second section will be devoted to modeling for the classic DTC control of the induction machine [7]. For this control strategy, we will explain the principle of electromagnetic flux and torque adjustment based on the discrete model of the two-level voltage inverter and using the generated voltage vectors. The proposed control model is presented and each block has been modeled [8]. The simulations results are presented to test the performances of the controller using Simulink.

The third section we define some strategies for improving DTC. Some intelligence techniques have been studied to improve the dynamic performance of the direct torque control, these methods consistent in replacing the conventional comparators and the selection table applied to the DTC of the induction machine with a controller based on the artificial intelligence by using the fuzzy logic.

2 Modelling of an Induction Machine

2.1 Construction

An induction machine is electrically a set of magnetically coupled windings. The stators of a synchronous machine and an induction machine are similar. They consist of three windings powered by three-phase quantities. These windings are distributed to create a resultant magnetomotive force, when energized by three-phase sinusoidal currents. The spatial distribution of this magnetomotive force is 120° degree as presented in Fig. 1 [9]. On the other hand, the rotor differs according to the type of machine. In the case of an asynchronous machine, the rotor has three windings

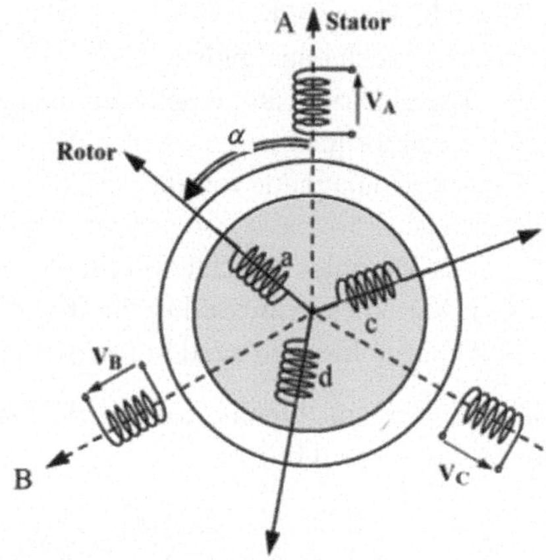

Fig. 1 Construction of an induction machine

distributed to have a sinusoidal resultant magnetomotive force whose mechanical period is 120° [10].

2.2 Mathematical Representation of an Induction Machine

The modeling of the induction machine is essential from the first step for its identification and its control. The mathematical model of the motor should, have such a structure to determine the characteristics of the motor [11]. We will use a model of the induction machine which describes the dynamic behavior of the different quantities concerned by the control system (electromagnetic torque, magnetic flux, currents, voltages, etc.). The equations of electrical motor are those of the flux and the voltage of the rotor and the stator [12, 13].

- Model Parameters
  Rs: Stator resistance.
  Rr: Rotor resistance.
  Ls: The self-inductance of a stator phase.
  Lr: The self-inductance of a rotor phase.
  Msr: Mutual inductance between stator and rotor phases.
  J: Moment of inertia of the rotor.
  P: Number of pole pairs.
- Machine Electrical and Mechanical Variables
  E: DC voltage at inverter input.
  Vs: The stator voltage.
  East: The stator current.
  Ir: The rotor current.
  Φs: The stator flux.
  Φr: The rotor flux.

Ws: The stator pulse.
W: The mechanical pulse.
θr: The electrical angle between the rotor and the stator.
Cr: Load torque.
Ce: electromagnetic torque.

- Clues

a, b, c: Variables indicated in the three-phase fixed frame.
d, q: Variables expressed in the frame (d,q) rotating at synchronous speed.
α, β: Variables registered in the two-phase fixed frame (α, β).

The induction machine is composed by six windings three in the stator and three in the rotor presented by:

$$\begin{aligned} V_{as} &= V_m*\cos(2\pi \text{ft}) \\ V_{bs} &= V_m*\cos\left(2\pi \text{ft - } \frac{2\pi}{3}\right) \\ V_{cs} &= V_m*\cos(2\pi \text{ft - } \frac{4\pi}{3} \end{aligned} \tag{1}$$

2.3 Electrical Equation

Choosing a model representation is made according to the type of control to achieve. Thus, for a motor supplied with voltage, the components of the equation of state control vector will tension [14]. The electrical equations of induction motor squirrel cage are those of flux and tension of the rotor and stator phases [15]. The voltage equations of the magnetically rotating stator and rotor circuits can be written in the following matrix form:

$$[V_r] = [R_r] * [I_r] + \frac{d[\varphi_r]}{\text{dt}}[V_s] = [R_s] * [I_s] + \frac{d[\varphi_s]}{\text{dt}} \tag{2}$$

The voltage equations of the stator and rotor can be written as:

$$
\begin{aligned}
V_{\mathrm{as}} &= R_s.I_{as} + \frac{d\varphi_{as}}{\mathrm{dt}} \\
V_{\mathrm{bs}} &= R_s.I_{bs} + \frac{d\varphi_{bs}}{\mathrm{dt}} \\
V_{\mathrm{cs}} &= R_s I_{cs} + \frac{d\varphi_{cs}}{\mathrm{dt}} \\
V_{\mathrm{ar}} &= R_r.I_{ar} + \frac{d\varphi_{ar}}{\mathrm{dt}} \\
V_{\mathrm{br}} &= R_r.I_{br} + \frac{d\varphi_{br}}{\mathrm{dt}} \\
V_{\mathrm{cr}} &= R_r I_{cr} + \frac{d\varphi_{cr}}{\mathrm{dt}}
\end{aligned}
\tag{3}
$$

This model is a system of six differential equations with periodic coefficient functions. The resolution is difficult. To overcome this problem, we use the two-axis three-phase voltage transformation. The conversion of a three-phase system (a, b, c) into a two-phase system clarke (d, q) is formalized as follows:

$$
\begin{aligned}
V_{ds} &= \sqrt{\frac{2}{3}}\left(V_{as} - \frac{1}{2}V_{bs} - \frac{1}{2}V_{cs}\right) \\
V_{qs} &= \sqrt{\frac{2}{3}}\left(V_{as} - \frac{\sqrt{3}}{2}V_{bs} - \frac{\sqrt{3}}{2}V_{cs}\right)
\end{aligned}
\tag{4}
$$

The two-phase voltages [Vds, Vqs] are input vectors and the currents [ids, iqs] are output vectors. The rotor voltage vector is zero because the rotor winding is shorted, Vqr = 0 and Vdr = 0.

$$
\begin{aligned}
V_{ds} &= R_s.I_{ds} - \omega_s.\varphi_{qs} + \frac{d\varphi_{ds}}{dt} \\
V_{qs} &= R_s.I_{qs} + \omega_s.\varphi_{ds} + \frac{d\varphi_{qs}}{dt} \\
V_{dr} &= R_r * I_{dr} - \omega_r.\varphi_{qr} + \frac{d\varphi_{dr}}{dt} = 0 \\
Vqr &= R_r.I_{qr} + \omega_r.\varphi_{dr} + \frac{d\varphi_{qr}}{dt} = 0
\end{aligned}
\tag{5}
$$

2.4 Magnetic Equation

The flux of the stator and rotor windings can be written in terms of winding inductances and current in reference (a, b, c) [16]:

$$\begin{aligned}[\varphi_s] &= [l_s].[I_s] + [M_{sr}].[I_r] \\ [\varphi_r] &= [l_r].[I_r] + [M_{sr}].[I_s]\end{aligned} \tag{6}$$

After long calculation Eqs. (1–6), the three to two axis transformations are applied to establish matrix between flux and currents:

$$\begin{bmatrix}\varphi_{ds}\\ \varphi_{qs}\\ \varphi_{os}\\ \varphi_{dr}\\ \varphi_{qr}\\ \varphi_{or}\end{bmatrix} = \begin{bmatrix} l_s - M_s & 0 & 0 & \frac{3}{2}.M_{sr} & 0 & 0\\ 0 & l_s - M_s & 0 & 0 & \frac{3}{2}.M_{sr} & 0\\ 0 & 0 & l_s + 2M_s & 0 & 0 & 0\\ \frac{3}{2}.M_{sr} & 0 & 0 & l_r - M_r & 0 & 0\\ 0 & \frac{3}{2}.M_{sr} & 0 & 0 & l_r - M_r & 0\\ 0 & 0 & 0 & 0 & 0 & l_r + 2M_r\end{bmatrix} . \begin{bmatrix}I_{ds}\\ I_{qs}\\ I_{os}\\ I_{dr}\\ I_{qr}\\ I_{or}\end{bmatrix} \tag{7}$$

Ls = ls-Ms: Stator inductance.

Lr = lr-Mr: Rotor inductance.

M = 3/2.Msr: Mutual inductance between rotor and stator.

Los = ls + 2 M: Inductor homopolar stator.

Lor = lr + 2 M: Inductor homopolar rotor.

If we eliminate the zero-sequence system (7) is obtained:

$$\begin{bmatrix}\varphi_{ds}\\ \varphi_{qs}\\ \varphi_{dr}\\ \varphi_{qr}\end{bmatrix} = \begin{bmatrix} L_s & 0 & M & 0\\ 0 & L_s & 0 & M\\ M & 0 & L_r & 0\\ 0 & M & 0 & L_r\end{bmatrix} . \begin{bmatrix}I_{ds}\\ I_{qs}\\ I_{dr}\\ I_{qr}\end{bmatrix} \tag{8}$$

By replacing the equations of flux in voltage equation:

$$\begin{bmatrix}V_{ds}\\ V_{qs}\\ V_{dr}\\ V_{qr}\end{bmatrix} = \begin{bmatrix} R_r + L_s.\frac{d}{dt} & -L.\frac{d\theta_s}{dt} & M.\frac{d}{dt} & M.\frac{d\theta_s}{dt}\\ L_s.\frac{d\theta_s}{dt} & R_s + L_s\frac{d}{dt} & L_s.\frac{d\theta_s}{dt} & M.\frac{d}{dt}\\ M.\frac{d}{dt} & -M.\frac{d\theta}{dt} & R_r + L_r\frac{d}{dt} & -L_r.\frac{d\theta}{dt}\\ M.\frac{d\theta}{dt} & M.\frac{d}{dt} & L_r.\frac{d\theta}{dt} & R_r + L_r\frac{d}{dt}\end{bmatrix} . \begin{bmatrix}I_{ds}\\ I_{qs}\\ I_{dr}\\ I_{qr}\end{bmatrix} \tag{9}$$

With the rotor in short circuit:

$$\begin{bmatrix}V_{ds}\\ V_{qs}\\ 0\\ 0\end{bmatrix} = \begin{bmatrix} R_s + L_s\frac{d}{dt} & -\omega_s.L_s & M.\frac{d}{dt} & \omega_s.M\\ \omega_s.L_s & R_s + L_s\frac{d}{dt} & \omega_s.L_s & M.\frac{d}{dt}\\ M.\frac{d}{dt} & -M.(\omega_s - \omega_r) & R_r + L_r\frac{d}{dt} & -L_r.(\omega_s - \omega_r)\\ M.(\omega_s - \omega_r) & M.\frac{d}{dt} & L_r.(\omega_s - \omega_r) & R_r + L_r\frac{d}{dt}\end{bmatrix} . \begin{bmatrix}I_{ds}\\ I_{qs}\\ I_{dr}\\ I_{qr}\end{bmatrix} \tag{10}$$

With:

$$\omega_s = \frac{d\theta_s}{dt}\omega_s - \omega_r = \frac{d\theta}{dt}$$

From the system of Eqs. (9–10) we can express the current in terms of flux as follows.

$$\begin{aligned} I_{ds} &= \frac{1}{\sigma.L_s}.\varphi_{ds} - \frac{1-\sigma}{\sigma.M}.\varphi_{dr} \\ I_{qs} &= \frac{1}{\sigma.L_s}.\varphi_{qs} - \frac{1-\sigma}{\sigma.M}.\varphi_{qr} \\ I_{dr} &= \frac{1}{\sigma.L_r}.\varphi_{dr} - \frac{1-\sigma}{\sigma.M}.\varphi_{ds} \\ I_{qr} &= \frac{1}{\sigma.L_r}.\varphi_{dr} - \frac{1-\sigma}{\sigma.M}.\varphi_{ds} \end{aligned} \tag{11}$$

With (σ: total leakage coefficient or scattering coefficient)

$$\sigma = 1 - \frac{M}{L_s.L_r}$$

2.5 Equations of Power and Torque

Conversions retain instantaneous power as:

$$\begin{aligned} P_i =& \left[R_s.I_{sd} + R._sI_{qs}\right] + \left[\frac{d\varphi_{ds}}{dt}.I_{ds} + \frac{d\varphi_{qs}}{dt}.I_{qs}\right] + \\ & \omega_s.\left[\varphi_{ds}.I_{qs} - \varphi_{qs}.I_{ds}\right] \end{aligned} \tag{12}$$

The first term represents the joule losses; the second term is the electromagnetic power; the third term corresponds to the electrical power. the electromagnetic torque Ce is given by:

$$\begin{aligned} P_e &= C_e.\Omega_s = \omega_s.\left(\varphi_{ds}.I_{qs} - \varphi_{qs}.I_{ds}\right) \\ C_e &= \frac{P_e}{\Omega_s} = \frac{\omega_s}{\Omega_s}.\left(\varphi_{ds}.I_{qs} - \varphi_{qs}.I_{ds}\right) \\ C_e &= P.\left(\varphi_{ds}.I_{qs} - \varphi_{qs}.I_{ds}\right) \\ C_e &= P.M.\left(I_{qs}.I_{qr} - I_{ds}.I_{qr}\right) \end{aligned} \tag{13}$$

2.6 Mechanical Equation

We call Ce the electromagnetic torque developed by the machine, Cr the resistive torque of the load, J the total inertia brought back to the shaft of the machine. The mechanical equation of induction machine is represented by [17].

$$J.\frac{d\Omega_r}{dt} = C_e - C_r \tag{14}$$

3 Model of an Induction Machine

3.1 Simulations

The simulation of an induction machine is presented in Fig. 2, we translate the mathematical model of the machine using the simulation blocks by the Matlab/Simulink. it contains the conversion, abc-dq, slip and efficiency blocks, Joule stator and rotor losses blocks.

In this schematic the simulation result is given by generating three-phase stator voltages and the transforming these voltages to two phase voltages. The current and electromagnetic flux equations were represented in this model.

The power of induction motor is 1.5KW with rated current, In = 3.5A and stator resistance is Rs = 6.72 Ω,, rotor resistance, Rr = 4.4 Ω, rotor inductance, Lr = 0.4614 H stator inductance Ls = 0.4246 H, mutual inductance M = 0.46H, with a speed NS = 1500 rpm and Cr = 25Nm.

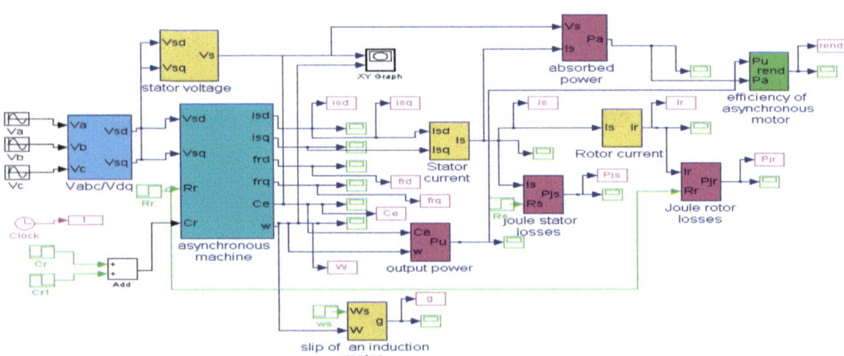

Fig. 2 Model of an induction machine

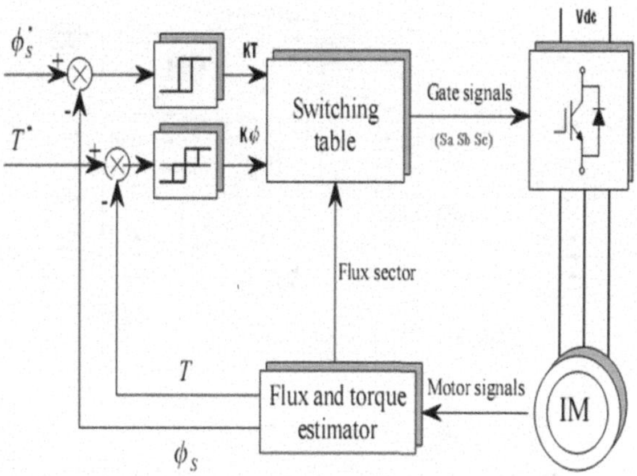

Fig. 3 DTC system of an induction motor

4 Control of an Induction Machine

4.1 Principles of Direct Torque Control of Induction Motor

The Direct Torque Control (DTC) method was proposed by Takahashi and Noguchi in 1986 [17]. This method is considered as a simple and robust method to control induction machines. The inverter represents the main element in this type of control.

The DTC method makes it possible to control two quantities which are the torque and the vector of the flux and electromagnetic. This control is done by determining the state of the inverter. The sinusoidal form of the flux makes it possible to obtain high performance.

The DTC consists in controlling the amplitudes of the flux and the torque using hysteresis comparators as in [17]. The Fig. 3 represents the principle of the DTC system. It includes several blocks: flux and torque hysteresis controllers, a commutation table and electromagnetic flux and torque estimators.

4.2 Direct Torque Control Model

The structure of the Direct Torque Control DTC of the induction motor in MATLAB/Simulink is shown in Fig. 4. The quantities obtained from the machine model $is\alpha$, $is\beta$, $Vs\alpha$ and $Vs\beta$ which are applied to a block of flux estimator and torque thus to the sector detector, these estimated values are compared with reference values obtained from the 2- and 3-level hysteresis correctors, consequently, its errors are integrated into a commutation table which is linked to a mains ratio to reduce ripples [18, 19].

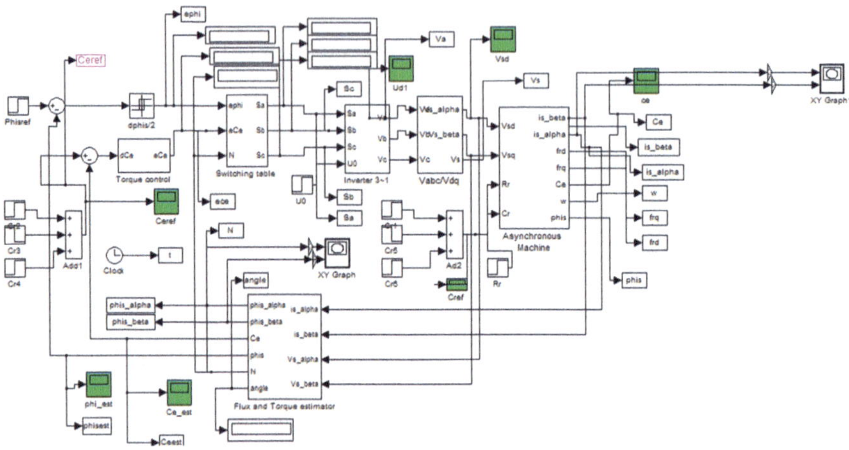

Fig. 4 Structure of DTC for induction machine

4.3 *Result of Simulation*

The induction machine is powered by an alternating voltage as shown in Fig. 5. Figure 6 shows that the electromagnetic torque varies at startup and after 0.02 stabilizes. The pace of the couple presents disturbance undulations.

The Torque and flux control is performed by the eight-voltage vector commutation table as shown in Fig. 7.

The Fig. 8 shows that the flux φs in the complex frame is circular. The Fig. 9

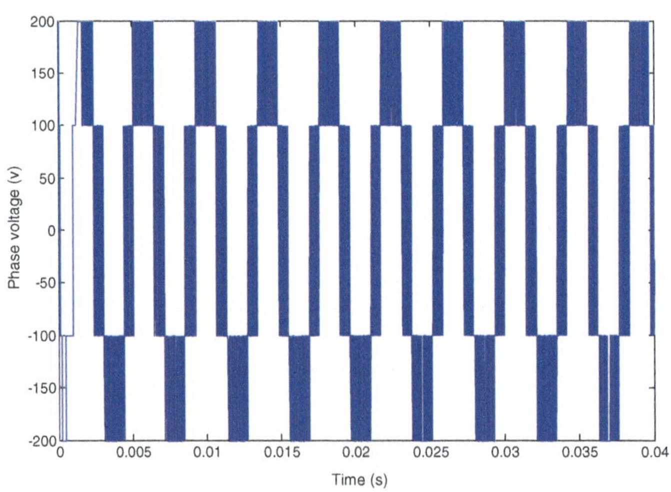

Fig. 5 Stator voltage

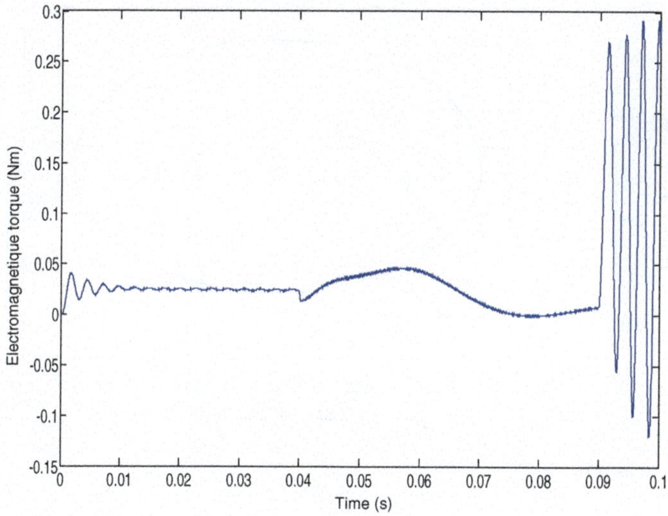

Fig. 6 Electromagnetic torque

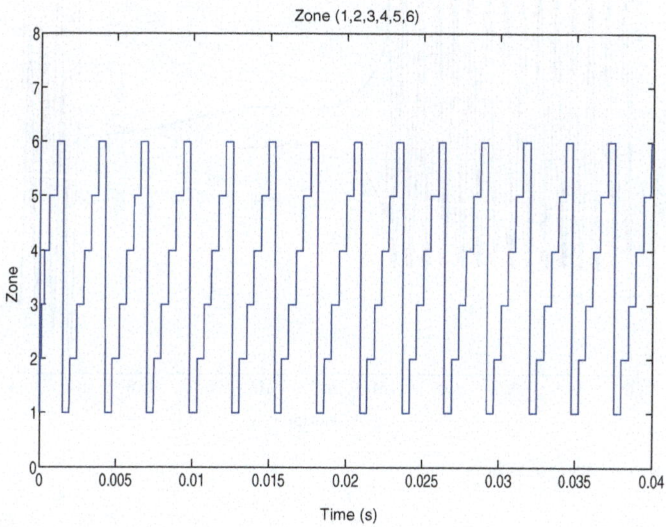

Fig. 7 Selector vector

shows that the flux variation leads to a significant variation in the stator current [20, 21].

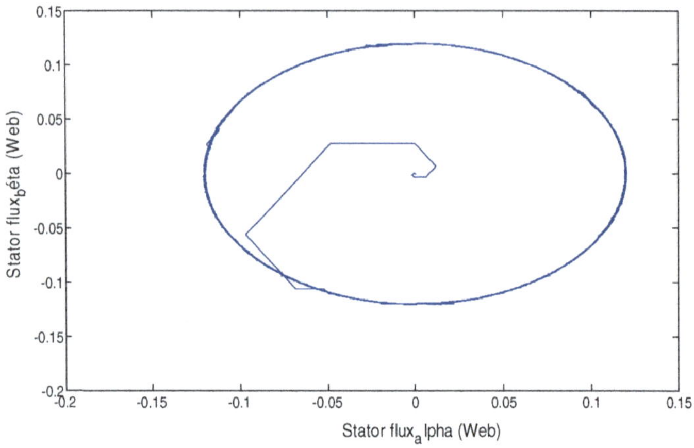

Fig. 8 Flux trajectory

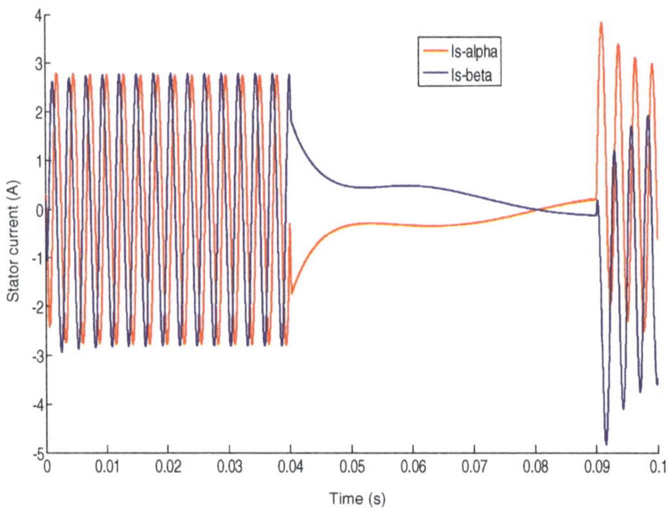

Fig. 9 Stator currents is α, is beta

5 Artificial Intelligent Control of an Induction Machine

5.1 Direct Torque Control by Fuzzy Logic

Fuzzy logic is a technique used in artificial intelligence. In fact, it had been around for a long time. It was Heisenberg's logical paradoxes and uncertainty principles that led to the development of "multiple-value logic" in the 1920s and 1930s. The

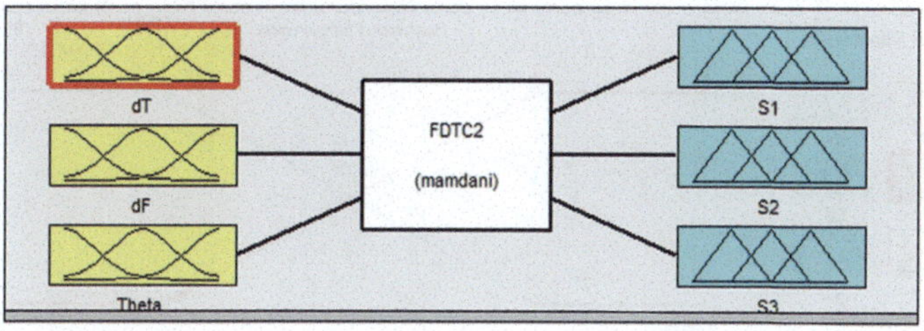

Fig. 10 Fuzzy controller

principle of fuzzy logic is like the human process such that the variables are linguistic variables.

The DTC control shows flux and torque ripples [22]. The technique of artificial intelligence based on fuzzy logic makes it possible to remedy the problem of ripples. this improvement consists in realizing a controller allowing to control the width of the band of hysteresis of the couple compared to the ripples of the couple. The type of fuzzy logic controller used in this method is a Mamdani type controller, which allows to establish a series of rules.

To test the performance of the fuzzy controller, we start with the fuzzification which consists in choosing the form of the membership function [23]. Then we present the rules of fuzzy reasoning in the IF–THEN format. The direct torque control system based on fuzzy logic consists of having three inputs and three outputs. The inputs are the flux and torque error and the stator angle the outputs are the switching state (S1, S2 and S3) allows to have the inverter voltages. The fuzzification part is implemented on Matlab Simulink as shown in Fig. 10.

5.2 *Error Flux*

The form of the membership function is of the trapezoidal type of which we present the flow error by two negative linguistic values noted N, positive noted P as shown in Fig. 11.

5.3 *The Torque Error*

The Eq. 15 shows that the torque error is the difference between the estimated torque value and the actual torque value. In this case we use a membership function composed of three fuzzy parts as shown in Fig. 12.

$$\Delta Ce = Ce * -Ce \quad (15)$$

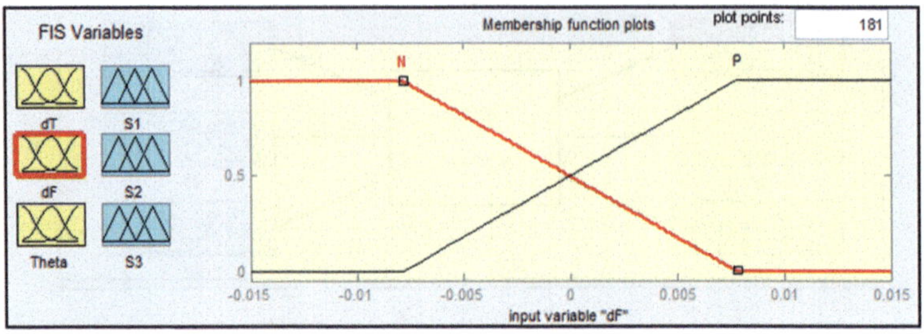

Fig. 11 Error membership functions flux

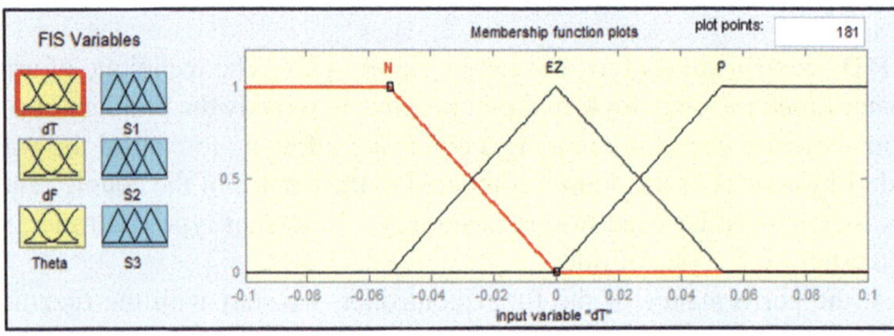

Fig. 12 Torque error membership functions control

5.4 *Flux Angle Θs*

The flux angle θs is defined by the angle between the flux Φs and the reference as shown in Eq. 16. Figure 13 presents the values of the flux angle.

$$\theta s = arctg(\frac{\varphi\beta s}{\varphi\alpha s})$$

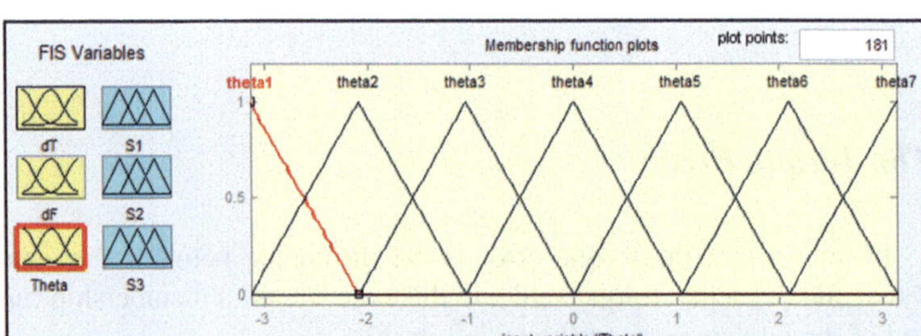

Fig. 13 Flux angle

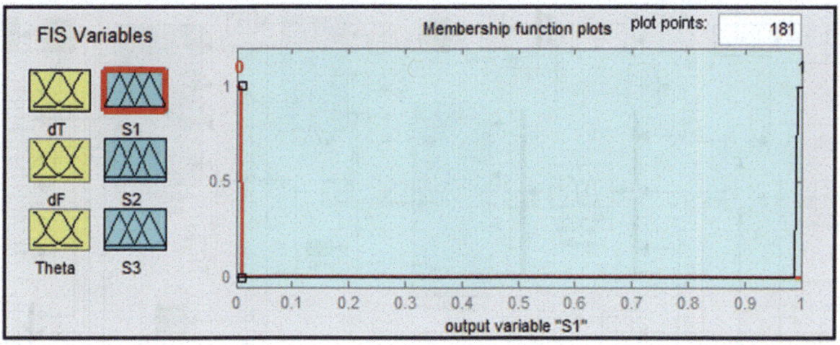

Fig. 14 Voltage membership functions

5.5 *Fuzzy Rules*

The main part in a fuzzy controller is rule-based. These rules are of the form IF and THEN are illustrated from the commutation table. We found 42 rules that match the DTC control. The fuzzy system includes many rule groups, each containing 42 rules.

5.6 *Voltage Membership Functions*

The classic DTC control makes it possible to obtain six voltage vectors to maintain the flux within the limit values of the hysteresis band [24]. In a fuzzy controller the form of the membership functions which corresponds to the output voltage vectors are shown in Fig. 14.

5.7 *Simulations Results*

Simulation In order to test the performance of DTC control using artificial intelligence based on fuzzy logic, a model of DTC fuzzy control was implemented and validated by the same 1.5 kW induction machine [25]. Figure 15 shows the model established by fuzzy logic.

The current delivered to the induction machine is very high at the moment of starting and it is of sinusoidal form as shown in Fig. 16. Figure 17 shows the shape of the supply voltage of the machine by a voltage converter.

The Fig. 18 presents the torque controlled with the fuzzy logic method based on artificial intelligence, we notice that the ripples are reduced compared to the response of the torque controlled with the classic DTC method.

The Figs. 19 and 20 show that the flux is equal to the reference value in classic DTC and fuzzy logic. The flux ripples in the fuzzy controller are decreased compared to the flux in the classic DTC.

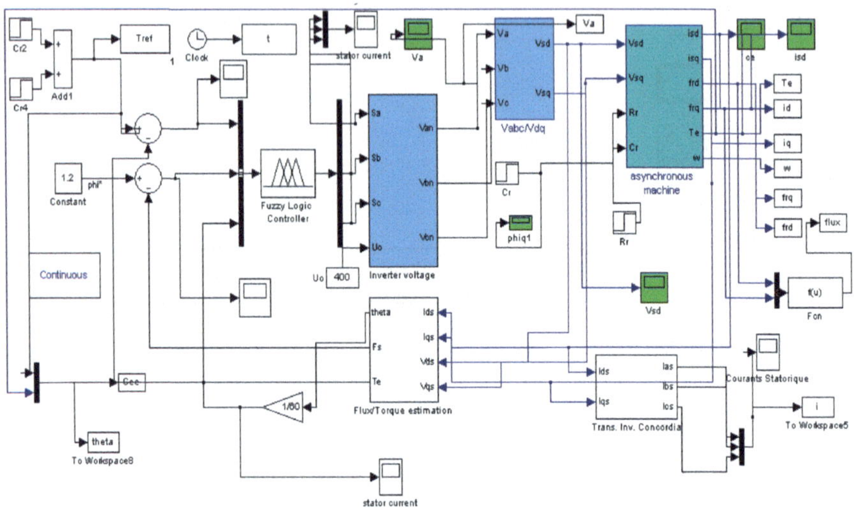

Fig. 15 Model of the fuzzy method

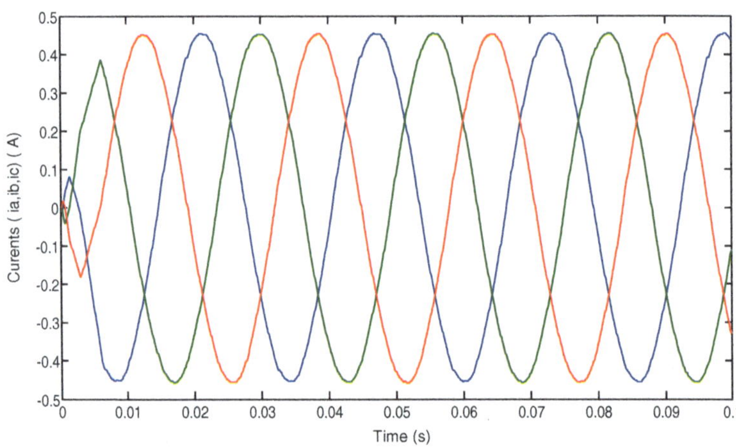

Fig. 16 Current with fuzzy method

6 Conclusion

In this chapter, we introduced the performance of DTC control of induction machine based on artificial intelligence. The model of this control was established using the modeling of the induction machine. This control technique makes it possible to improve and remedy robustness problems.

This type of control allows the torque and flux of the machine to be controlled without the need for a mechanical sensor. The simulations gave satisfactory results in

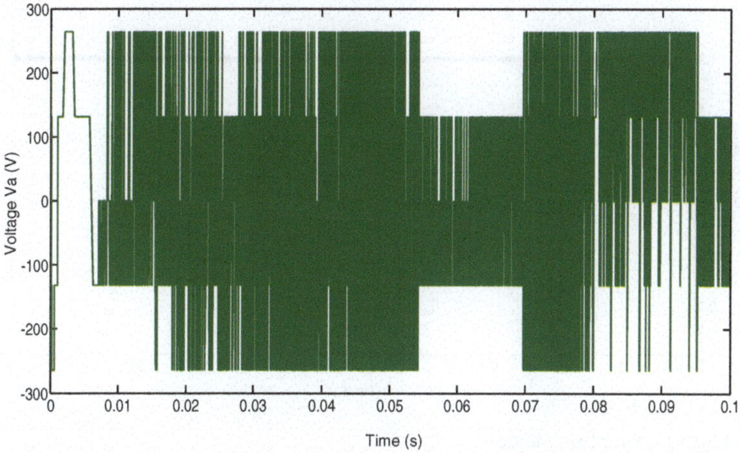

Fig. 17 Stator voltage with fuzzy logic

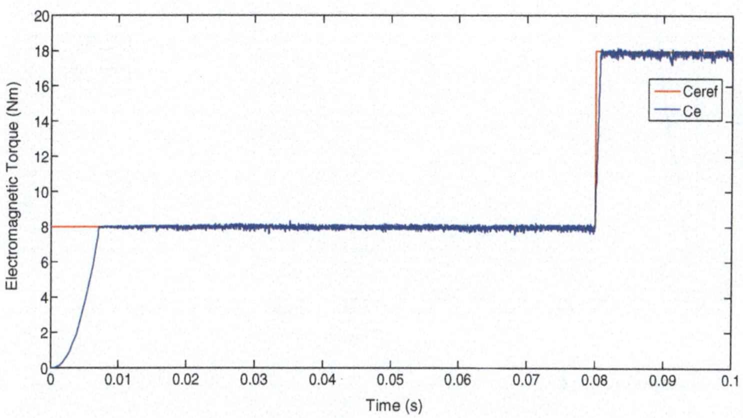

Fig. 18 Torque response with fuzzy logic

torque and flux control. This control also makes it possible to reduce the current at the start of the induction machine, which protects the protection devices of the machine. However, this control has some drawbacks such as high ripple of torque, flux, current and variations in the switching frequency of the inverter. To minimize torque ripples in classic DTC control. The control technique using artificial intelligence based on fuzzy logic has been implemented in Matlab Simulink. Motor torque ripples have decreased. At the same time, flux fluctuations are reduced compared to the classic DTC control technique. the hysteresis control and the switching table used in the classic method are eliminated.

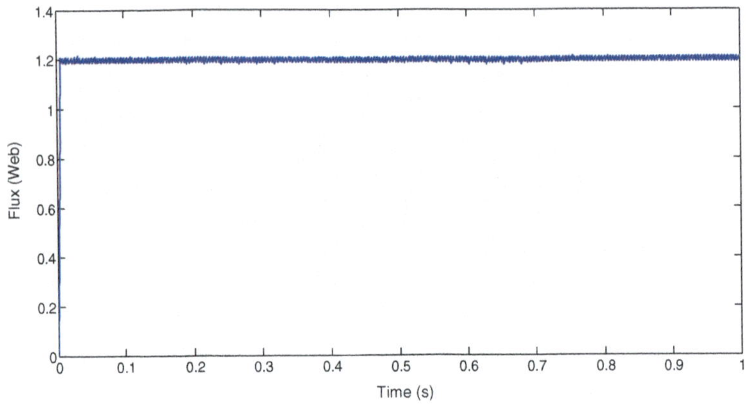

Fig. 19 Flux estimator fuzzy logic

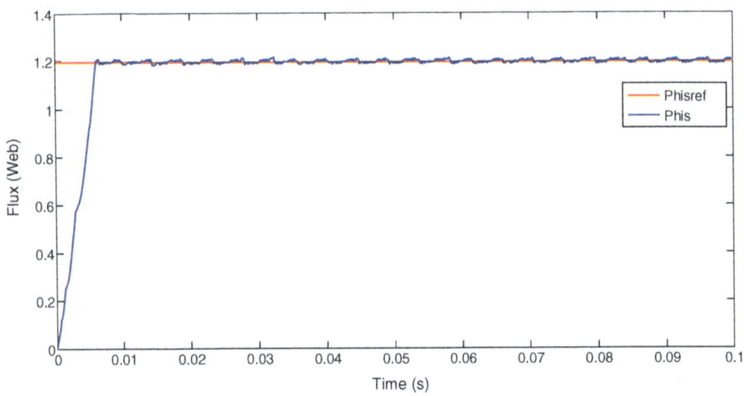

Fig. 20 Flux estimator DTC

References

1. Takahashi, I., Noguchi, T.: Take a look back upon the past decade of direct torque control. In: IECON 23rd International Conference, vol 2, issue, 9–14, pp. 546–551 (1997)
2. Depenbrock, M.: Direct self-control (DSC) of inverter-fed induction Machine. IEEE Trans. Power Electron. **13**(2) (1988)
3. Mekrini, Z., Bri, S.: Performance of an indirect field-oriented control for induction machine. J. Eng. Technol. (IJET SCOPUS Index. J.) **8**(2), 726–733 (2016)
4. Mekrini, Z., Bri, S.: Fuzzy logic application for intelligent control of an induction machine. Indones. J. Electr. Eng. Comput. Sci. (IJEECS SCOPUS Index. J.) **7**(1), 61–70 (2017)
5. Zadeh, L.A.: Fuzzy sets. Inf. Control **8**, 338–353 (1965)
6. Jain, A.K., Mao, J., Mohiuddin, K.M.: Artificial neural networks: a tutorial. IEEE Mag.: Comput. **29**(3), 31–44 (1996)
7. Wai, R.-J., Lee, J.-D., Lin, K.-M.: Robust decoupled control of direct field-oriented induction motor drive industrial electronics. IEEE Trans **52**(3), 837–854 (2005)

8. Mensou, S., Essadki, A., Nasser, T., Idrissi, B.B.: A direct power control of a DFIG basedWECS during symmetrical voltage dips. Prot. Control. Mod. Power Syst. (Springer), pp 1–12 (2020)
9. Mensou, S., Essadki, A., Minka, I., Nasser, T., Idrissi, B.B., Bentarla, L.: Performance of a vector control for DFIG driven by wind turbine: real time simulation using DS1104 controller board. Int. J. Power Electron. Drive Syst. **10**(2), 1003–1013 (2019)
10. El Ouanjli, N., Motahhir, S., Derouich, A., El Ghzizal, A., Chebabhi, A., Taoussi, M.: Improved DTC strategy of doubly fed induction motor using fuzzy logic controller. Energy Rep. **5**, 271–279 (2019)
11. Mensou, S., Essadki, A., Nasser, T., Idrissi, B.B., Tarla, L.B.: Dspace DS1104 implementation of a robust nonlinear controller applied for DFIG driven by wind turbine. Renew. Energy Elsevier **147**, 1759–1771 (2020)
12. Mensou, S., Essadki, A., Minka, I., Nasser, T., Idrissi, B.B. : Control and hardware simulation of a doubly fed induction aero-generator using dSPACE card. In: Proceeding of the 2nd International Conference of Computer Science and Renewable Energies (2019)
13. Rahaman, N., Govindraju, H.V.: Modeling & simulation of a three-phase electric traction induction motor using matlab simulink. Int. J. Electr. Electron. Comput. Syst. (IJEECS) **2**(5, 6), 18–25 (2014)
14. Ayasun, S., Nwankpa, C.: Induction motor tests using MATLAB/Simulink and their integration into undergraduate electric machinery courses. IEEE Trans. Educ. **48**(1) (2005)
15. Anbarasu, E., Karthikeyan, M.: Modeling of induction motor and fault analysi. Int. J. Eng. Sci. Innov. Technol. (IJESIT) **2**(4) (2013)
16. Arabaci, H., Bilgin, O.: Squirrel cage of induction motors simulation via Simulink. Int. J. Model. Optim. **2**(3), 324–327 (2012)
17. Issac, S., Vanamathi, K.: Modelling of single Sided Linear Induction Motor by MATLAB/SIMULINK. Int. J. Soft Comput. Eng. **2**(3), 492–494 (2012)
18. Palpankar, P.M., Ghanmare, R.U., Makade, N.: A generalized dynamic model of induction motor using simulink. Trans. Electr. Electron. Eng. (ITSI-TEEE)32 **1**(5), 118–122 (2013)
19. Wieczorek, M., Rosołowski, E.: Modelling of induction motor for simulation of internal faults. Mod. Electr. Power Syst. **3**(9), 29–35 (2010)
20. Nwodo, T.C.: Simulation tools for electrical machines modelling: teaching and research. Niger. J. Technol. **24**(1), 33–37 (2015)
21. Jain, S.K., Sharma, F., Baliwal, M.K.: Modeling and simulation of an induction motor. Int. J. Eng. Res. Dev. **10**(4), 57–61 (2014)
22. Singh, R., Bajpai, S., Sandhu, H.S.: Comparative study of PWM control and PI control of induction motor. Bull. Electr. Eng. Inf. **4**(1), 53–58 (2015)
23. Mohan, K.S, Febin Daya, JL: Machine model based speed estimation schemes for speed encoderless induction motor drives: a survey. Bull. Electr. Eng. Inf. **4**(1), 7–17 (2015)
24. Mekrini, Z., Bri, S.: High-performance using neural networks in direct torque control for induction machine. Int. J. Electr. Comput. Eng. (IJECE a SCOPUS indexed Journal) **8**(2), 31–38 (2018)
25. Alnasir, Z.A., Almarhoon, A.H.: Design of direct torque controller of induction motor (DTC). Int. J. Eng. Technol. (IJET) **4**(2), 54–70 (2012)

Evaluation of the Recommended Algorithms in the Internet of Things

Emad R. Gadalla, Omar M. Sallabi, Tahani M. Kasih, and Abdulrahman A. A. Emhemed

Abstract In the recent revolution of exchanging information, more and more objects (things) become in need of a media that allows them to talk to each other. Perhaps, the Internet is, so far, the most suitable infrastructure to establish a connection between billions of things. Internet of Things (IoT) phrase was introduced to express this fact. The internet security challenges have always been the concern of scientists. However, IoT imposes new demands since the connected things (Devices) usually have limited capabilities. Therefore, different types of security solutions are needed to tackle such demands. Our work is with the authentication security phase, which has issues associated with the IoT structure nature. One of the major issues in this field is the authentication issue. In this paper, we discussed the authentication issues associated with IoT, and then a proposed authentication solution is provided. Our concern is to construct an abstract end-user authentication approach that considers IoT requirements. According to our results, security solutions for IoT have to utilize algorithms, which have good performance.

Keywords Internet of things · Lightweight cryptography · Measurement

E. R. Gadalla (✉)
Faculty of Engineering Sciences, Bright Star University, El-Brega, Libya
e-mail: Emad.Rabeh@bsu.edu.ly

O. M. Sallabi · T. M. Kasih
Faculty of Information Technology, University of Benghazi, Benghazi, Libya
e-mail: omar.sallabi@uob.edu.ly

T. M. Kasih
e-mail: tahini.kasih@uob.edu.ly

A. A. A. Emhemed
Faculty of Technical Engineering, Bright Star University, El-Brega, Libya

S. El Himer et al. (eds.), *Artificial Intelligence of Things for Smart Green Energy Management*, Studies in Systems, Decision and Control 446,
https://doi.org/10.1007/978-3-031-04851-7_9

1 Introduction

A great deal of different small devices that recently connected to the Internet (such as actuate, sensors, mobile phones, etc.) is known as IoT, which imposes new network-security-related challenges. Some works have dealt with IoT issues before. However, in [1], Kevin Ashton is the first one who came up with the Internet of Things in 1999. The IoT technically presents many connected physical objects that provide various services using the Internet. Due to the remarkable development of Information Communication Technologies (ICT), many exclusive services (such as telemedicine, e-health, smart home applications, … etc.) become available for individuals through the IoT networks that are usually heterogeneous.

Nevertheless, people expect to use and connect these services securely; no clear-cut solution guarantees this. Information and knowledge are often exchanged via worldwide networks (mainly the Internet) in this era, which is a fantastic way of sharing data. Unfortunately, it left unauthorized parties with the considerable possibility of accessing these data (because of the infrastructure nature of the Internet). As a result, designing a mechanism to protect such data while transferring has been within the researcher's interests. In the IoT, the concern falls into the following issues [2]:

- Authentication: The job is to build a protocol assuring that the communicating entity is the party it claims to be.
- Access control: Here, the work is developing a mechanism that limits and controls access to host systems.
- Attack resistant: It is about detecting unauthorized parties and preventing their attacks.

Security of IoT is based on providing the authentication and authorization of IoT nodes (things, users, servers, objects), data authenticity, confidentiality, integrity. The security solutions are mostly performed in the following layers: network, transport, and application. To secure first two levels, the following security protocols can be used: IPSec, Transport Layer Security (TLS) protocol, Host Identity Protocol (HIP), Datagram Transport Layer Security (DTLS) protocol, and Slim Extensible Authentication Protocol Over Local Area Networks (SEAPOL). Privacy-preserving solutions are based on cryptographic normal [3]. Cryptography is the science of mathematical algorithms for ensuring confidentiality, integrity, and authenticity of the information. Cryptography algorithms can be symmetric and asymmetric. Public-key encryption is called for asymmetric encryption, which utilizes a pair of keys—a public key for encrypting data and a private key—for decrypting data. As the mentioned name, Symmetric encryption means that the encryption and decryption operations use the same key. However, traditional cryptography does not guarantee an optimal level of security. Furthermore, existing security solutions in IoT are still vulnerable to such attacks: Denial of Service, Man-in-Middle, Eavesdropping.

2 Related Works

Devices in IoT usually have limited capability to store data (Low-Capacity memory). They are also, in most cases, strangers to each other, i.e., there is no previous handshake. Therefore, Identification, authentication, and access control are pre-requested processes to establish secure communication. However, the frontier fact, the efficient solutions for these processes have to be lightweight [4] to consume memory that does not affect the device's performance. This leads to adopting solutions that mitigate the cryptographic schemas [5] utilised in the authentication phase since their parameters (such as keying material) usually occupy a considerable chunk of memory. On the other hand, such solutions would allow attackers to illegally access data stored in the IoT devices.

In the first, security challenges related to IoT have lately been at the top of the specialized priorities list. In [6, 7] MAGNET project presents a close work that deals with the denial of services (DoS). However, authentication issues are not considered.

This project proposed an IoT-suited access control approach based on the user's credential; nevertheless, attack resistance is left unaddressed. In [5, 8], authentication protocol, which is dependent upon Elliptical Curve Cryptography (ECC), is presented, but there is no mention of how to tackle the Denial of Service (DoS) attack. Since there are many systems in IoT, it is essential to sort out the problems caused by the (DoS) attack. In [9], problems with securing communication and achieving authentication using shared keys are addressed. The proposed approach can only be used within a limited location, and there is no room for an extension to a wide area. The authentication part is not adaptable to work in an IoT environment.

Researchers are divided on which cryptographic primitives satisfy the IoT requirements, public key or secret key. Lots of authentication techniques were developed based on secret-key cryptographic technologies, and this is because of its fast computation. Unfortunately, due to memory needed for storing keys, it is not suitable for devices in IoT that usually have low-capacity memory.

In contrast, solutions based on public-key cryptography do not suffer from such issues because they require low memory, and there is no need to exchange the secret key to establish the first communication. In [4], the ECC-based mutual authentication approach for IoT is introduced using hash functions (electronic fingerprint). Handshake is accomplished between terminal nodes. The secret-key cryptography-based platform is designed to address the problems with storage and key management. An authentication scheme that uses certified keys is presented in [10].

This system is only used to authenticate user nodes, and it is not suitable for IoT because it is not a lightweight solution. Handshake in [11] is performed using authentication with a parameter. This process is based on secret-key cryptography, which requires more memory, and is time-consuming. An authentication protocol based on a cluster is introduced in [12]. This could be the most suited solution for IoT. However, it is vulnerable to attackers who can grab the keying material and use them to lead a DoS attack.

Few papers investigate security in IoT, including cryptography techniques. For instance, Sicari et al. [13] give a diagram of the heterogeneous technologies in IoT. In such a situation, remarkable advances and correspondence benchmarks, which ought to give security and protection prerequisites, are absent. Appropriate arrangements must ensure protection for clients and things, control, secrecy, and reliability among clients and gadgets. This work proposes open issues for the future explores the mix of IoT and correspondence innovations in a safe middleware; IoT security in cell phones.

Sklavos and Zaharakis [14] tackled alternative technologies and devices in the Internet of things, which relied on a combination of different models, schemas, and application aspects. Their work focused on the efficiency of hardware implementation, confidentiality of communications, user authentication, data integrity, and availability of services. They also look at recent attacks, threats, and countermeasures against them. For general purposes, such a system was introduced: To ensure confidentiality, they use the asymmetric algorithm (RSA), the Advanced Encryption Standard (AES), and digital signatures to manage the keys. Elliptic Curve Cryptography (ECC) and Diffie-Hellman (DH) can be applied as an alternative to RSA. Formulated by Katagi and Moriai [15], it showed the meaning of light encryption. According to this definition, cryptography is a cryptographic account or protocol designed to be implemented in restricted environments, including RFID tags, sensors, smart cards without contact, medical services, etc. This paper outlines the state of institutional filing for cryptography and explains why lightweight encryption is required for IoT. Usman et al. [16] propose lightweight cryptographic algorithms—Secure IoT (SIT).

The algorithm combines Feistel and a uniform substitution-permutation network with 64-bit block encryption and a 64-bit key. The result showed that this algorithm is suitable for a low-cost 8-bit microcontroller due to code size, memory usage, and implementation time of encryption/decryption. Things can be adopted in Internet applications but still need to be tested on different platforms for possible attacks. Nandhini and Vanitha [17]. Present a review of lightweight modern coding in their research work [13]. They suggest a new branch of lightweight encryption, such as comparing low resource requirements and the performance and strength of encoded primitives. In this research, an analysis was done for the modern lightweight blade such as HISEC, OLBCA, PRINCE, PRINT, PRESENT, KLEIN, TWINE. Some of these algorithms use the Feistel network, others are SPN, but each has its characteristics. Analysis showed that if the algorithm contains many S boxes, security will be good, but the cost will be high. On the other hand, if the algorithm has enough blocks, S and A sophisticated linear process will provide high security and cost-dependent design.

This related work presents the definition of light encryption and explains its mechanisms. Comparison between different algorithms is also provided because of the size of the code, the use of memory, and the time of implementation of encryption/decryption. Results are based on low computational devices, including 8-bit microcontrollers. However, they do not consider any specific scenario in Internet things and suggest no solution to the scenario and the research gap mentioned above to be considered.

3 Design and Implementation

The importance of identification authentication processes is presented. A use-case is given to illustrate a real-life scenario for using the identity of things. Based on this research, the steps for obtaining our goal will be explained in detail. First, underlying hardware technologies (microcontroller) used for this study will be given. Second, we elaborate on various lightweight algorithms utilized and implemented for performance comparisons. Then, the recommendation of the scheme is explained in succeeding subsections. Finally, measurement metrics used for measuring performance are given.

3.1 Use Case

Here, a use case in the context of using thing identity in a real-life scenario is defined. The 70-year-old client Christina has made a few IDs on her cell phones (smartphones, tablets, laptops, etc.) and uses them to access various services (applications, platforms, web services, etc.). Because of her age, Christina takes her health and has an electronic health platform installed in her home. The platform includes restorative and kinetic sensors that can be worn and supported by ambient sensors. The e-health platform provides Christine's remote health monitoring service that helps them live independently. If necessary, it informs the carers concerned and their dependents to access their information with a specific end goal for correct activities. Christina visits her to focus on the movement from month to month and spends seven days there. It brings its tools but unfortunately cannot take home sensors. Some common equipment is also used in the center. When you want Cristine to access your local Wi-Fi network, you request your Wi-Fi password for the Activity Center at Classic mode. After that, you must write it on every device you want to use. In addition, they should activate many services that support their subsidized lives.

Conversely, in our case, all of their identities are stored in a web cloud (where the Wi-Fi password is stored in the activity center) and can easily be used after verifying its identity. Validation usually requires that you retype your user name and password. Fortunately, Cristine uses a smart identity management system that offers automatic user identification to find mobile devices. Kristen is automatically registered, and all of her devices will have seamless Wi-Fi access. Behind this process, Identity Management automatically activates its mapping of services and enables it to collect information in a meaningful personal way. Thus, the system provides response services regardless of time and place—Fig. 1 is displayed in a Use-case diagram.

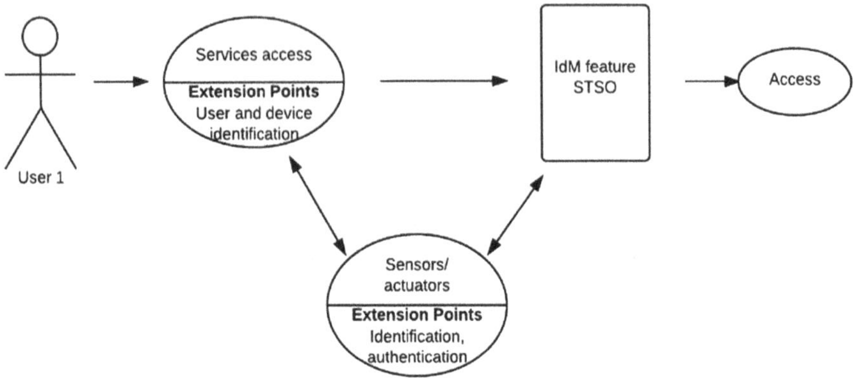

Fig. 1 Use-case diagram

3.2 *Wireless Sensor Networks and Internet-of-Things*

WSN networks consist of many smart, visual, and brilliant tools for radio detection. Developments in equipment, programming, and systems management have made WSN feasible, feasible, and feasible in many areas, such as land control, welfare, structure control, precision farming, and military.

The ongoing scan in the Internet zone has been accepted into the ID (for example, IPv6 addresses) of the installed tools installed within the current Web base on the use of the assets. In this way, tools grown on the Internet are things that can be addressed globally and speak of agreements on the Internet, for example, 6LoWPAN. The Internet encourages machine-to-machine (M2M) correspondence without human mediation and enables paid applications, for example, home control and wearable control. Center around WSNs, especially in lightweight arrangements, because of the assets of restricted equipment for the tools used. Nitty–gritty data about prerequisites, conventions, and engineering is displayed in [18]. A brief overview of the device resources is provided below to justify the proposed solution.

As was said by the few tools used, in this way, the devices have limited resources that give devices a name: "Restricted Devices." They are constrained in memory, arithmetic limits, and energy. RFC 722 [19] has identified and categorized effective tools in classes in the light of memory assets. These devices can be grouped into the following three categories:

- **Class 0** devices are axes similar to sensors, most pre-designed and have less than 10 kilobytes of RAM respectively 100 KB flash. Everything is said in Class 0 devices, and it is not ready to connect specifically and securely to the Web. Therefore, to participate in Web correspondence, the largest tools are required. Consider the field of projects Class 0 machines can not establish or supervise correspondence in the customary sense.
- **Class 1** devices have about ten kByte RAM and 100 kByte Flash available. Separately 100 kByte flash accessible. Unlike Category 0 tools, they can not effectively

talk to other Web hubs using a full stack of conventions (HTTP, TLS, security conventions, and XML-based information). For the most part, Class 1 tools use particularly expensive groups (for example, CoAP via UDP). In this way, you do not need hub hubs for discussion purposes. Class 1 devices can assist the security capabilities required on large systems, as is the scope of this theory. In addition, these tools can be formatted as a fully integrated aspect of IP regulation.

- **Class 2** devices are more extravagant memory, mostly containing about 50 KB RAM respectively 100 kByte Flash available. Furthermore, 250 kByte Streak is accessible. They can support most of the same convention interventions used on platforms or note servers with this range. However, class 2 devices can, in any case, benefit from lightweight and dynamic production agreements with a specific goal for building a lifetime. These tools contrast with other tools at a smaller level than their systems management assets, allowing access to more applications. In addition, if Class 2 devices, in addition, could enhance convention interventions to reduce the costs of improving the classroom, the interoperability could also be expanded. The tools with the basic capabilities of category two must be said in the past, given that they have also been integrated into the proper diversity of devices in Internet objects. These instruments can often use existing conventions in an unchangeable manner, but restricted vital supplies may now bind them. Mobile phones are a case for these Class 2+ devices.

All tools within Class 0 and Class 1 collect "Less Hardware" aggregation; for all tools in Category 2 or higher, the expression "Full Resource Tool" is used.

4 Limited IoT Device

Arduino is selected as a microcontroller device for our study to implement different lightweight algorithms. Arduino is an open-source electronics platform based on easy-to-use hardware and software. Arduino processors can be programmed by running code that will manage the device according to a given algorithm. Programs for Arduino can be written in C/C++, supplemented with simple and understandable functions for managing I/O operations. There is a free official programming environment, "Arduino IDE," running on Windows, Mac OS, and Linux for the convenience of working with Arduino. Loading a new program into Arduino becomes a matter of one click; connect the card to the computer via USB. Although for more inquiring minds, it is possible to work through Visual Studio, Eclipse, other IDEs, or the command line. Figure 2 shows the official programming environment "Arduino IDE." The comparison analysis of performances is based on the implementation of algorithms on the Arduino platform MEGA 2560. The Mega 2560 is a microcontroller board with 54 digital input/output pins, 16 analog inputs, 4 UARTs (hardware serial ports), and 16 MHz crystal oscillators. It has 256 KB of flash memory for storing code (8 KB is used for the bootloader), 8 KB of SRAM, and 4 KB of EEPROM. The Mega 2560 can be powered via a USB connection or an external power supply. The

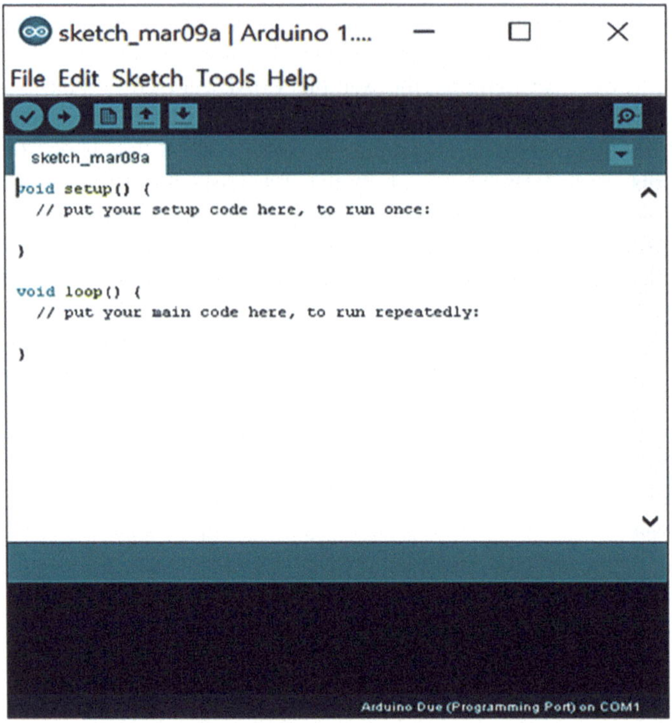

Fig. 2 Example of the official programming environment "Arduino IDE"

power source is selected automatically [20]. Figure 3 shows an example of shown pinout diagram of microcontroller Arduino MEGA.

5 Lightweight Algorithms

Lightweight cryptography is the collection of cryptographic primitives, techniques, and ciphers that can be applied in devices that cannot provide most of the existing codes and have limited resources (memory, power, size) for the operation. The main criteria for lightweight ciphers is a balance between reliability, performance, and price (see Fig. 4). The reason for this eternal search is requirements for algorithms: productivity, safety, and cost. It is easy to optimize two of them in practice: safety and productivity, cost and performance, or safety and cost, but it is still difficult to provide all requirements. That is why there are many implementations of lightweight cryptography both as software and hardware. Figure 4 shows the Interaction of reliability, performance, and price scheme. This work presents the comparative analysis of a few symmetric algorithms. They analyze based on the comparison of the performance of each algorithm (encryption/decryption time, memory, and power usage).

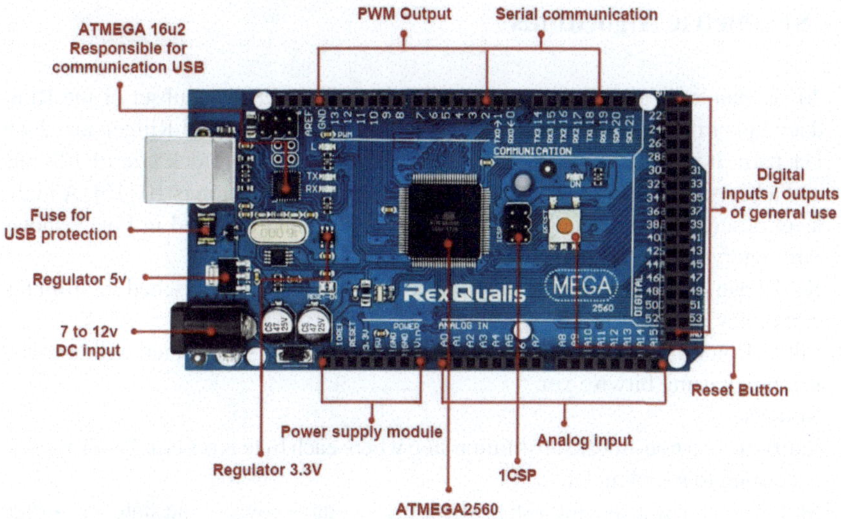

Fig. 3 Arduino mega pin description

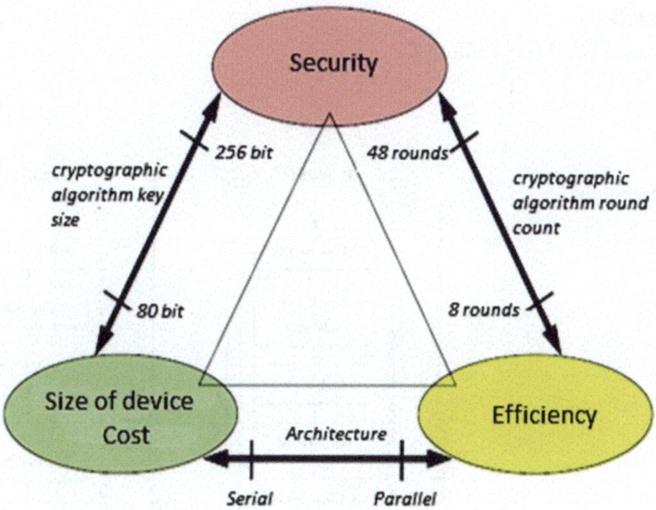

Fig. 4 Interaction of reliability, performance, and price scheme

For the comparison, three symmetric algorithms are chosen: AES, CHACHA, and SPACK. The choice is based on the latest research works, which propose to apply these algorithms for ensuring the security of IoT devices [5, 9, 10, 15].

6 Symmetric Algorithms

- **AES**: Standards and Technology (NIST) in 2001. AES is a subset of the Rijndael cipher developed by two Belgian cryptographers, Vincent Rijmen and Joan Daemen. It is an asymmetric encryption algorithm with a block size of 128 bits and a key size of 128 bits. The number of cycles of repetition is 10 [16]. A high-level description of the algorithm is given below, and depicted in Fig. 5 shows AES encryption/decryption algorithm.
- KeyExpansions—round keys are derived using Rijndael's key schedule from the cipher key.
- Initial Round. AddRoundKey—each byte of the state is combined with a round key block using bitwise xor.
- Rounds.
- SubBytes—a non-linear substitution step where each byte is replaced with another according to a lookup table.
- ShiftRows—a transposition step where the last three rows of the state are shifted a certain number of steps cyclically.
- MixColumns—a mixing operation that operates on the columns of the state, combining the four bytes in each column.
- AddRoundKey.
- Final Round (no MixColumns).
- SubBytes.

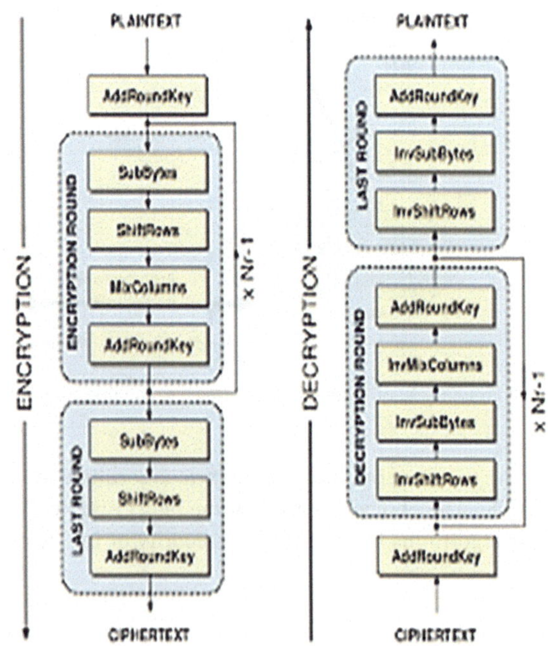

Fig. 5 AES encryption/decryption algorithm

- ShiftRows.
- AddRoundKey.

- **SPECK**: is a class of lightweight block ciphers publicly released by the National Security Agency (NSA) in 2013. Beaulieu et al. [21] Speck has been enhanced for algorithms and software implementations, while Simon algorithms have been optimized for hardware implementations. Speck is an add-rotate-xor (ARX) cipher. The NSA began working on the Simon and Speck ciphers in 2012. The agency anticipated that some agencies in the US federal government would need a cipher that would operate well on a diverse collection of Internet of Things devices while maintaining an acceptable level of security. [21] Speck is highly optimized for software implementation on microcontrollers. This class implements the Speck family that uses 128-bit block sizes with 128-bit, 192-bit, or 256-bit key sizes. Other Speck families support smaller block sizes of 32, 48, 64, or 96 bits, but such block sizes are too small for use in modern cryptosystems. The SpeckTiny and SpeckSmall types give alternative applications to reduce RAM and flash size requirements at the cost of some achievements and features. A high-level description of the algorithm is given below, and depicted in Fig. 6 shows SPCK encryption/decryption algorithm.

- **CHACHA**: the ChaCha family of stream ciphers, also known as Snuffle 2008, is a variant of the Salsa20 family of stream ciphers. It modifies the Salsa design to obtain a better performance and increased diffusion. ChaCha follows the same basic design principles as Salsa20, but I changed some of the details, most importantly to increase the amount of diffusion per round. The ChaCha stream cipher forms the basis of the finalist to the SHA-3 hash function competition. The ChaCha cipher is used within the web browser Chrome. A ChaCha has 16 Xors and 16 additions and 16 constant distance rotations of 32-bit words, like a Salsa20 round. Also, ChaCha has the same levels of performance and similarity as Salsa20 and saves one of the 17 registers used by a "natural" Salsa20 implementation. So it is reasonable to guess that a ChaCha round can achieve the same software speed as a Salsa20 round—and even better speed than a Salsa20 round on some platforms.
- Consequently, if ChaCha has the same minimum number of secure rounds as Salsa20, ChaCha will provide better overall speed than Salsa20 for the same level of security. ChaCha is a stream cipher that takes a key, an 8byte nonce/IV, and a counter and hashes them to generate a key stream to XOR with the plain text. Variations on the ChaCha cipher use 8, 12, or 20 rounds of hashing operations with 128-bit or 256-bit keys. A high-level description of the algorithm is given below and depicted in Algorithm: CHACHA quarter round functions.
- Input: Key K, Counter C, and Nonce N
- Output: Keystream Z
- Generate initial matrix X using K, C, and N.

Fig. 6 SPRCK encryption/decryption algorithm

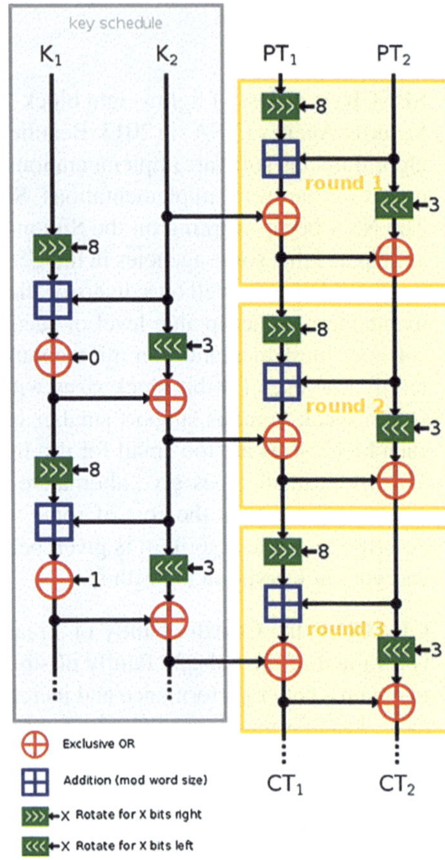

$y \rightarrow X fori \rightarrow 0to9do$
$/ * ColumnRound * /$
$(x0, x4, x8, x12) \rightarrow quarterround(x0, x4, x8, x12)$
$(x5, x9, x13, x1) \rightarrow quarterround(x5, x9, x13, x1)$
$(x10, x14, x2, x6) \rightarrow quarterround(x10, x14, x2, x6)$
$(x15, x3, x7, x11) \rightarrow quarterround(x15, x3, x7, x11)$
$/ * DiagonalRound * /$
$(x0, x5, x10, x15) \rightarrow quarterround(x0, x5, x10, x15)$
$(x1, x6, x11, x12) \rightarrow quarterround(x1, x6, x11, x12)$
$(x2, x7, x8, x13) \rightarrow quarterround(x2, x7, x8, x13)$
$(x3, x4, x9, x14) \rightarrow quarterround(x3, x4, x9, x14)$
$endfor$
$Z \rightarrow X + y$
$returnZ$

A vector (a, b, c, d) of four words is transformed; Fig. 7 shows the CHACHA encryption/decryption algorithm.

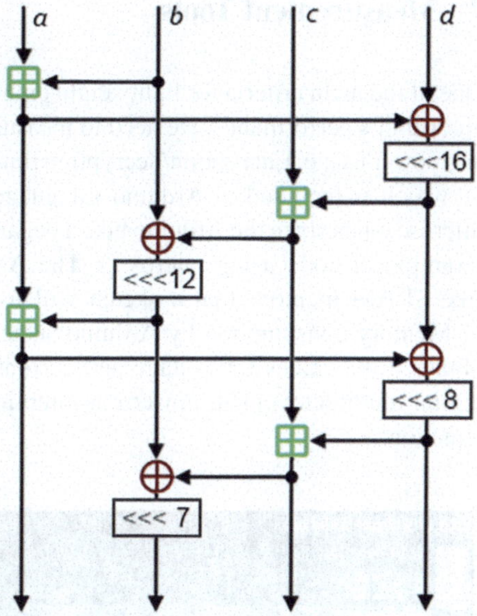

Fig. 7 CHACHA encryption/decryption algorithm

The algorithm math of Fig. 7 shows CHACHA encryption/decryption algorithm and shows the vector (a, b, c, d) of four words is described as:

$$
\begin{aligned}
&a = a + b \\
&d = d \oplus a \\
&d = (d) <<< 16 \\
&c = c + d \\
&b = b \oplus c \\
&b = (b) <<< 12 \\
&a = a + b \\
&d = d \oplus a \\
&d = (d) <<< 8 \\
&c = c + d \\
&b = b \oplus c \\
&b = (b) <<< 7
\end{aligned}
$$

The quarter-round functions are applied to the columns (x0, x4, x8, x12), (x5, x9, x13, x1), (x10, x14, x2, x6) and (x15, x3, x7, x11) in odd round, and diagonals (x0, x5, x10, x15), (x1, x6, x11, x12), (x2, x7, x8, x13) and (x3, x 4, x9, x14) in even rounds. Algorithm describes the complete procedure of CHACHA.

7 Measurement Tools

One of the main criteria for lightweight ciphers is their performance. To measure the algorithm's performance, we need to measure encryption/decryption time, memory, and power usage. Encryption/decryption time can be measured using function micros (), which is included in Arduino Language. This function returns the number of microseconds since the Arduino board began running the current program see Fig. 8. Example of code using micros (). The Arduino IDE shows both the amount of free SRAM memory that a sketch will use after compiling see Fig. 9 Example of Memory consumption by Arduino sketch. To measure power consumption, we obtained the values for voltage and current from the METRAHit 16S—analog to digital multimeters [16]. Power consumption P was estimated using the following equation:

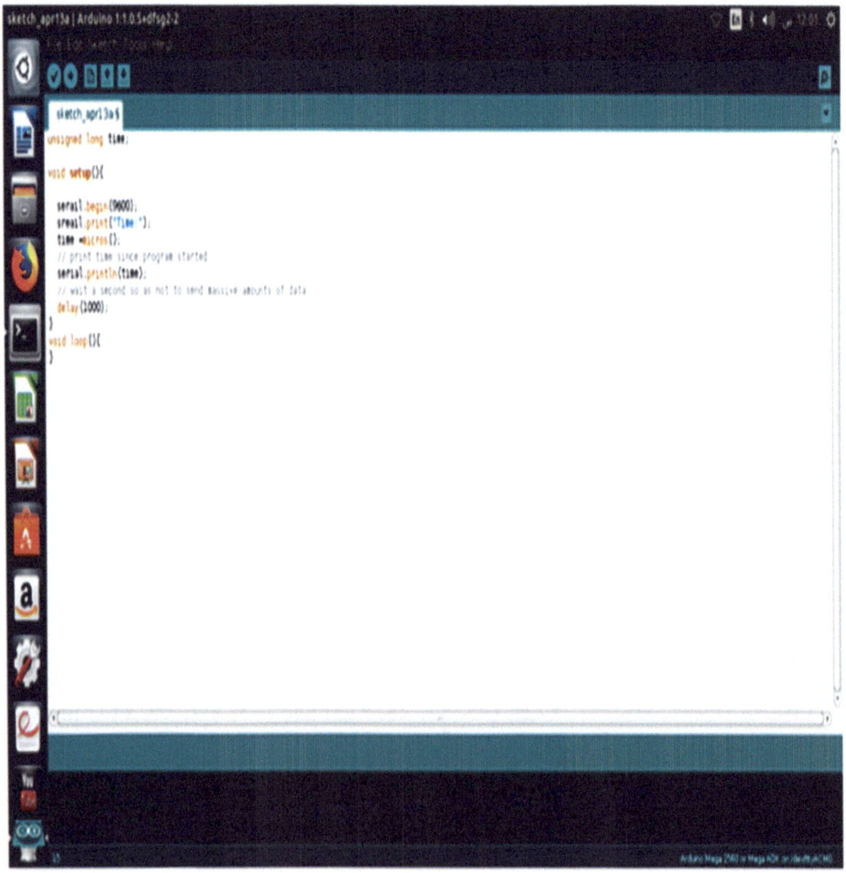

Fig. 8 Example of code using micros ()

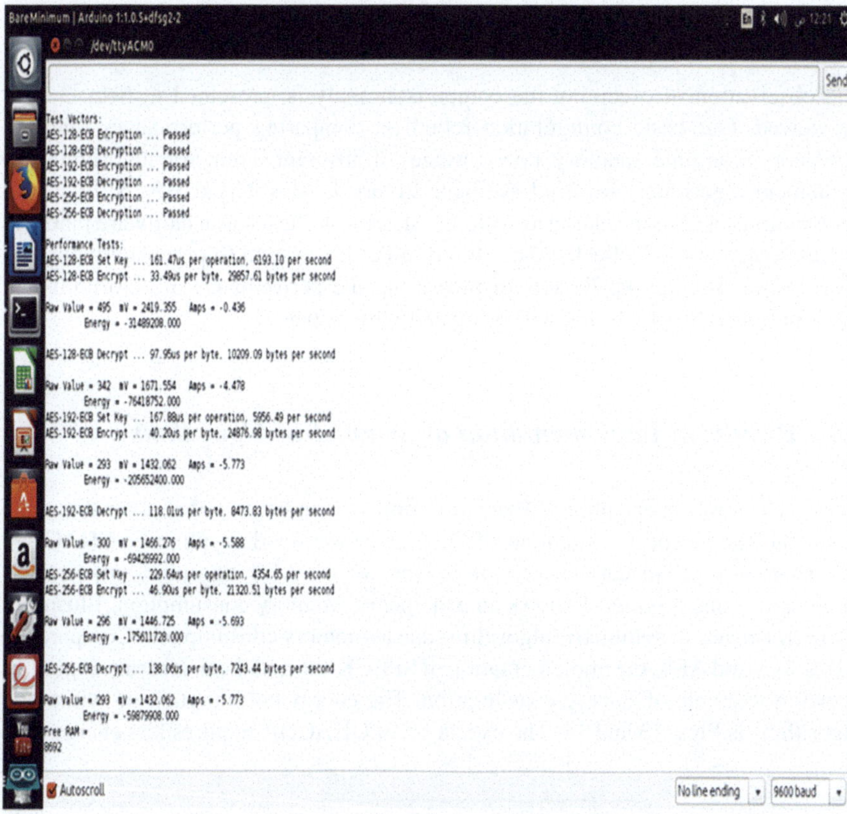

Fig. 9 Example of memory consumption by Arduino sketch

$$P = V * I$$

where V and I refer to voltage and current, respectively. To measure energy consumption from the previous equation, we obtained the value for power consumption.

$$E = P * T$$
$$E = V * I * T$$

8 Results and Analysis

The observation of executing the comparison analysis, presented in Sect. 3, will be viewed. Our basic configuration relied on comparing performances (encryption/decryption time, memory, power usage) of different symmetric algorithms. To implement algorithms, we used available libraries: AES, CHACHA, SPECK. An open-source electronics platform Arduino Mega was chosen as a hardware platform for implementation. Collected data stored in the file system to produce graphs and analyze the observation. Results of measuring the performance of algorithms are used for generating plots and will be given in this paper.

8.1 Results of Implementation of Symmetric Algorithms

Figure 10 shows A example of Execution time. It can be seen that algorithm AES has a shorter execution time than SPECK and time for decryption of CHACHA. The shortest time for data encryption belongs to the AES algorithm among these three algorithms. Figure 11 shows an example of Memory consumption. Illustrates the performance of symmetric algorithms due to memory consumption. Compared to CHACHA and AES, the implementation of SPECK showed the best result. Figure 12 shows A example of Energy consumption. The energy consumption of symmetric algorithms is Figs. 13 and 14. The execution of CHACHA requires less energy than

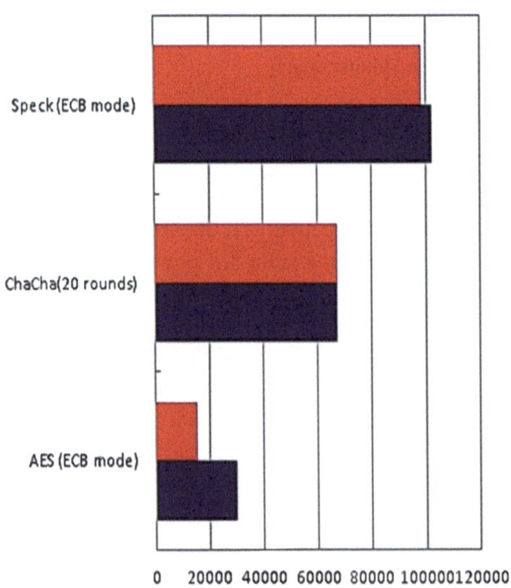

Fig. 10 Example of execution time

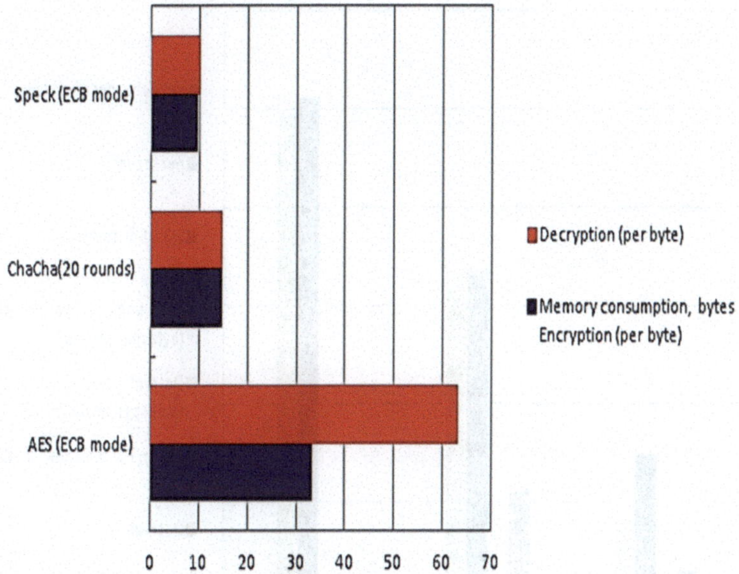

Fig. 11 Example of memory consumption

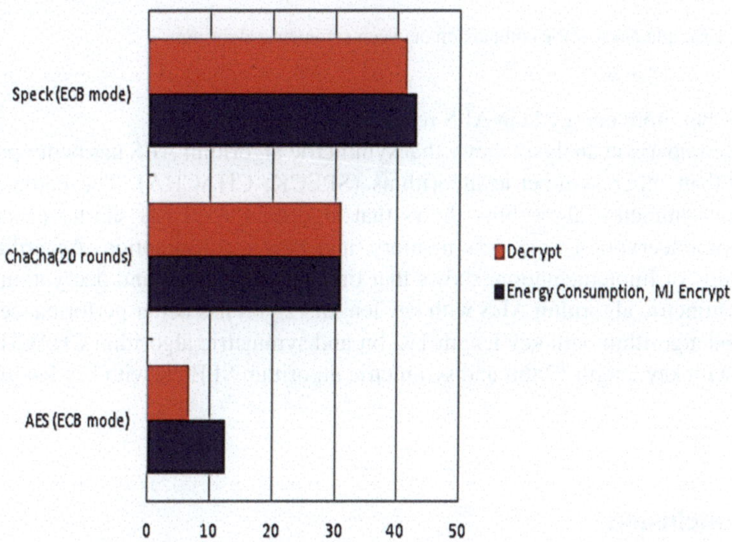

Fig. 12 Example of energy consumption

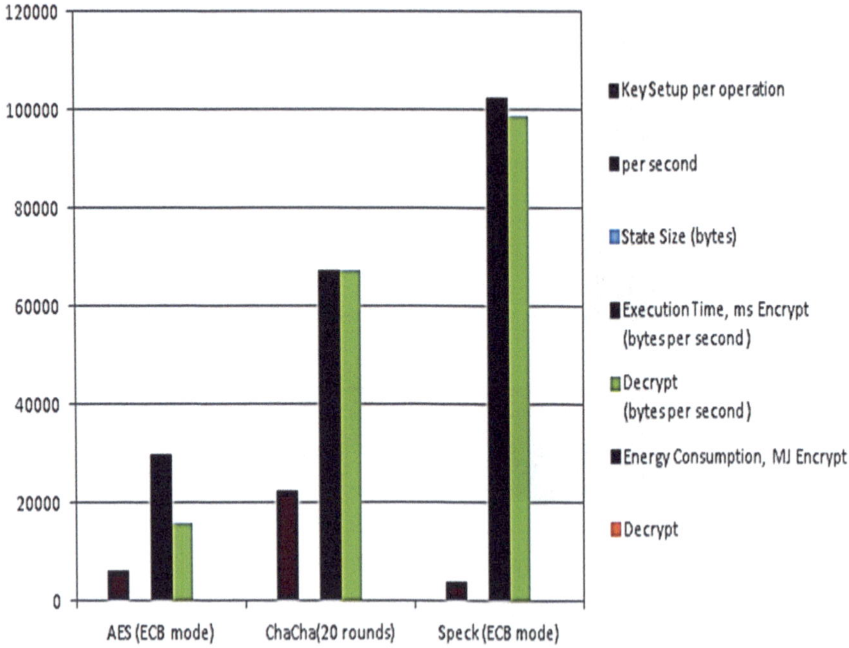

Fig. 13 Example No.1 of the comparison between symmetric algorithms

SPECK but more energy than AES regarding data encryption.

The comparison analysis shows that symmetric algorithm AES has better performance than other symmetric algorithms (SPECK, CHACHA). The comparison between symmetric algorithms shows that algorithm AES has shorter execution encryption/decryption time, less memory, and power consumption. According to the results of implementation, shows that time for encryption and decryption data with symmetric algorithm AES with key length 128 bit has better performance than the same algorithm with key length 192 bit and symmetric algorithm CHACHA 12 round with key length 128bit and symmetric algorithm SPECK with key length 192 bit.

9 Conclusion

This research proposed a method to provide cryptographic security for IoT devices. Because of the inherent limitations of the devices used in IoT in terms of processing power, memory, and energy, it is not easy to implement sufficient cryptographic functions. Therefore, lightweight cryptography was a recommendation to solve this problem.

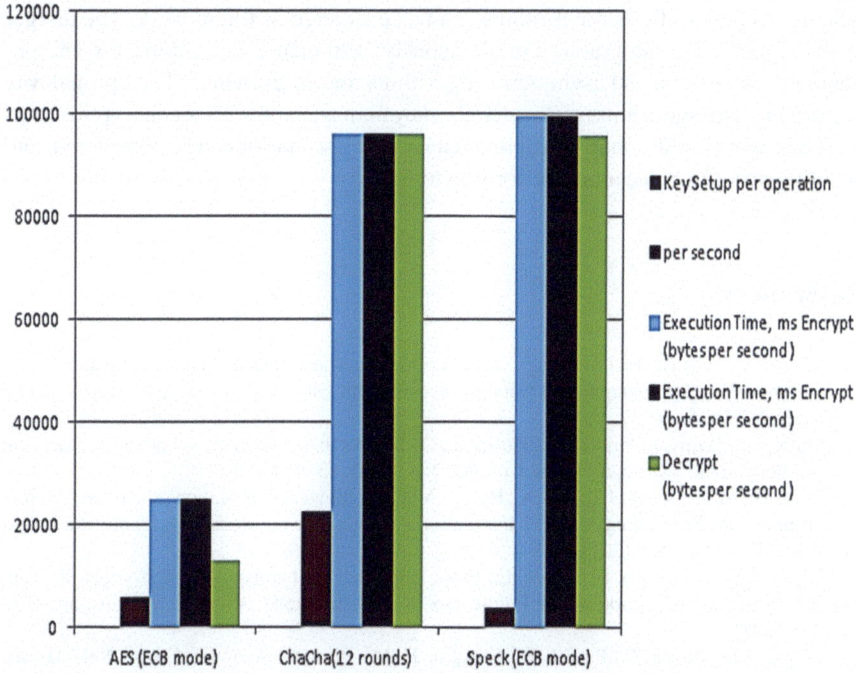

Fig. 14 Example No.2 of the comparison between symmetric algorithms

Lightweight cryptography is a section of cryptography that aims to develop algorithms for use in devices with limited recourses (memory, power, etc.). We propose the optimal cryptography solution for IoT devices. Since symmetric algorithms provide confidentiality, integrity, authenticity, and asymmetric algorithms provide all security requirements. However, their key size is too large, making them more complex and unsuitable for applying in IoT; we recommend using an AES algorithm, which can be employed in systems with better energy consumption. Low-cost devices can be applied in every IoT sector when data for sending is small. For instance, it can be blood pressure, heart rate, breath rate, and blood sugar in the medical sector. In the field of agriculture, this scheme can send data such as temperature, light level, soil moisture, PH levels, etc.

10 Future Work

This research is based on only the implementation of a few algorithms. This research is based on only the implementation of a few algorithms, which is limited compared with other algorithms. and no mention of the performances of others. Also, we used only one type of electronics platform for our comparative analysis and testing of the scheme. Analysis of other algorithms can be considered as future work. The merger of two types of cryptography simultaneously: symmetric algorithms for encryption/decryption data; and asymmetric algorithms for key exchange. The optimal way to combine two algorithms is to use ECC as key manager and AES for encryption data. As future work. Additionally, different data sending scenarios can be considered, and new schemes can be designed according to them.

References

1. Marsan, C.: The Internet of Things: An Overview. Internet Society, Tech. Rep. (2015)
2. Stallings, W.: Cryptography and Network Security, 4th edn. Prentice-Hall Inc., Upper Saddle River, NJ, USA (2005)
3. Malina, L., Hajny, J., Fujdiak, R., Hosek, J.: On perspective of security and privacy-preserving solutions in the internet of things. Comput. Netw. **102**, 83–95 (2016)
4. Zhao, G., Si, X., Wang, J., Long, X., Hu, T.: A novel mutual authentication scheme for internet of things. In: Proceedings of 2011 International Conference on Modelling, Identification and Control, pp. 563–566 (2011)
5. Ahamed, S.I., Rahman, F., Hoque, E.: Erap: Ecc based Rfid authentication protocol. In: 12th IEEE International Workshop on Future Trends of Distributed Computing Systems, pp. 219–225 (2008)
6. Olsen, R.L., Fitzek, F., Prasad, R., Saarnio, J.: My Personal Adaptive Global Net-Magnet, WWRF8bis. Beijing, China, pp. 26–27 (2004)
7. Kyriazanos, D.M., Stassinopoulos, G.I., Prasad, N.R.: Ubiquitous access control and policy management in personal networks. In: Third annual international conference on mobile and ubiquitous systems: networking services, pp. 1–6 (2006)
8. Braun, M., Hess, E., Meyer, B.: Using elliptic curves on Rfid tags. Int. J. Comput. Sci. Netw. Secur. **8** (2008)
9. Balfanz, D., Smetters, D.K., Stewart, P., Wong, H.C.: Talking to strangers: authentication in ad-hoc wireless networks. In: NDSS. Citeseer (2002)
10. Du, W., Wang, R., Ning, P.: An efficient scheme for authenticating public keys in sensor networks. In: Proceedings of the 6th ACM International Symposium on Mobile Ad Hoc Networking and Computing. ACM, pp. 58–67 (2005)
11. Verma, R.R.S., O'Mahony, D., Tewari, H.: Progressive authentication in ad hoc networks. In: Proceedings of the Fifth European Wireless Conference (2004)
12. Venkatraman, L., Agrawal, D.P.: A novel authentication scheme for ad hoc networks. In: 2000 IEEE Wireless Communications and Networking Conference. Conference Record (Cat. No.00TH8540), vol. 3, pp. 1268–1273 (2000)
13. Sicari, S., Rizzardi, A., Grieco, L.A., Coen-Porisini, A.: Security, privacy and trust in the internet of things: the road ahead. Comput. Netw. **76**, 146–164 (2015)
14. Sklavos, N., Zaharakis, I.D.: Cryptography and security in the internet of things (IoT): models, schemes, and implementations. In: New Technologies, Mobility and Security (NTMS), 2016 8th IFIP International Conference on. IEEE, pp. 1–2 (2016)
15. Katagi, M., Moriai, S.: Lightweight Cryptography for the Internet of Things. Sony Corporation, pp. 7–10 (2008)
16. Usman, M., Ahmed, I., Aslam, M.I., Khan, S., Shah, U.A.: Sit: A Lightweight Encryption Algorithm for Secure Internet of Things (2017). arXiv preprint arXiv:1704.08688
17. Nandhini, P., Vanitha, V.: A study of lightweight cryptographic algorithms for IoT. **6**, 26–35 (2017)
18. Karl, H., Willig, A.: Protocols and Architectures for Wireless Sensor Networks. Wiley (2007)

19. Bormann, C., Ersue, M., Keranen, A.: Terminology for constrained node networks. Tech. Rep. (2014)
20. Luhach A.K., et al.: Analysis of lightweight cryptographic solutions for the internet of things. Indian J. Sci. Technol. **9**(28) (2016)
21. Beaulieu, R., Shors, D., Smith, J., Treatman-Clark, Weeks, S., B., and Wingers, L.: The Simon and Speck families of lightweight block ciphers, cryptology e-print archive. Indian J. Sci. Technol. (2013)

Exposing Applications of IoT in Green Computing

Keshav Kaushik

Abstract Internet of Things (IoT) offers global networking of diverse IoT-enabled devices to exchange relevant information, thanks to scientific and technological advances. Tremendous energy needs arise from the frequent interchange of a massive volume of data via the IoT network. The primary goal of green IoT is to alleviate the challenges that come with maintaining a sustainable ecosystem. It has been noted that the sector of IoT-green computing necessitates several important enhancements with large startup costs. It is not readily accessible to everyone in society because of a low level of awareness and technical competence. This chapter highlights the introduction to IoT and green computing by discussing the primary applications of IoT in green computing. Moreover, the challenges and issues related to IoT and green computing are also discussed in this chapter.

Keywords e-waste · Internet of Things · IoT · Green computing · Power management · Smart technologies

1 Introduction

Green computing, sometimes referred to as green technology, is the practice of using computers and other information technology supplies and systems in an energy-efficient and environmentally responsible manner. Green computing organizations often utilize energy-efficient central processing units (CPUs), servers, accessories, and transmission networks. They also place a premium on conserving resources and appropriately disposing of both physical and technological trash. Green computing refers to reducing the environmental effect of technology. This comprises energy conservation, waste minimization, and long-term sustainability. Greening technology has the ability to increase ecological responsibility for making the entire product life cycle of inventions and things greener, including research, production, use, and disposal.

K. Kaushik (✉)
School of Computer Science, University of Petroleum and Energy Studies, Dehradun, India
e-mail: officialkeshavkaushik@gmail.com

S. El Himer et al. (eds.), *Artificial Intelligence of Things for Smart Green Energy Management*, Studies in Systems, Decision and Control 446,
https://doi.org/10.1007/978-3-031-04851-7_10

Green computing's main goal is to minimize energy usage. This not only saves money on electricity for businesses, but it also minimizes the carbon footprint of IT systems. In the research of data center design, the environmental effect of IT components is critical. Computers and other computational power have become very energy-efficient assets because to improvements in energy development and preservation. To make data centers, office buildings, and other high-energy assets more ecologically sustainable, green design has become a crucial aspect of new construction and infrastructure improvements.

In all technology advancement [1], such as Big Data and the Internet of Things, the Green Computing concept promotes energy-proportional and efficient computing resource consumption (IoT). IoT allows for global connection across physical things all over the globe. People cooperating automatically with heterogeneous gadgets has made the world smarter as research and technology has progressed. Massive amounts of data must be sent between devices, which necessitates a lot of energy. The Internet of Things is linked to big data and analytics and cloud computing to foresee smart device behavior, provide crucial business insights, and serve as a feedback control. Additionally, most businesses would adapt to the growing interest in the digital world, resulting in a quicker growth in energy usage.

Furthermore, the quantity of producers and users of various technologies is increasing at an exponential pace. Green IoT [2] is a subclass of the Internet of Things that seeks to minimize environmental challenges and maintaining a long-term environment. It has been emphasised in several IoT deployments that deploying energy-efficient equipment either reduces or limits the greenhouse impact. The Green IoT helps to maintain weather conditions by introducing low-energy devices or electrical appliances, reducing greenhouse gas emissions, using carbon-free materials, and fostering reusability. A summary of Green IoT and its applications is presented with the purpose of minimizing IoT energy usage in order to build a sustainable environment.

Data centers, equipment rooms, storage facilities, and other components that utilize or are influenced by energy consumption are often the focus of IT managers' energy efficiency initiatives. One of the motivating forces is the desire to save money. Green initiatives are also aided by government restrictions relating to energy saving. A third reason driving the green movement is concern about global warming, as well as personal and societal pressures to be ecologically responsible. Green computing follows several tactics in the IoT industry to make the working efficient and provide ecofriendly environment. Figure 1 shows the green computing tactics in IT industry.

- **Use of Smart Technologies**—To gather and interpret information about the data center and construct a power use model, enterprises may employ internet of things devices and artificial intelligence (AI) surveillance technologies. In the data center, AI-powered gadgets can also handle heating, cooling, and electricity on their own.
- **Working Remotely**—The COVID-19 epidemic has prompted several adjustments in the workplace, including some that have resulted in lower energy use. The number of persons traveling to and from work has reduced. It has also reduced

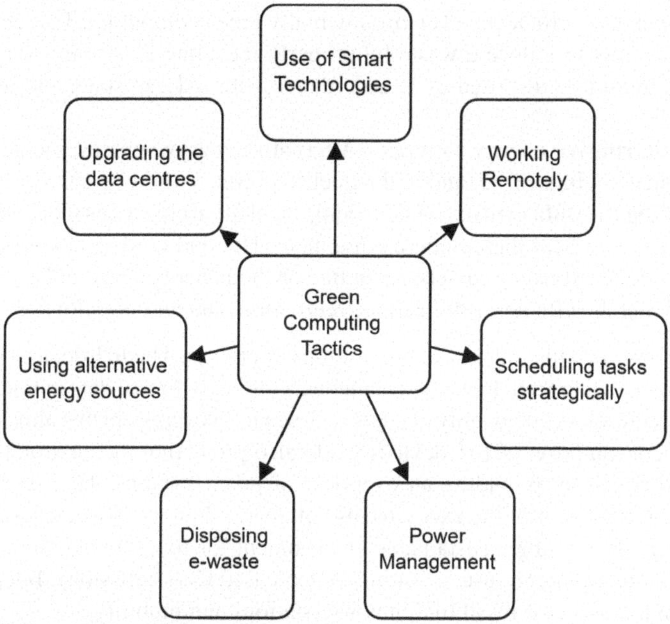

Fig. 1 Green computing tactics in IoT

the number of staff working in a company's facilities, lowering consumption on energy, water, and other commodities.

- **Scheduling tasks strategically**—Do computer-related work in blocks of time, turning off hardware while not in use. Laptops utilize a fraction of the energy that desktop PCs do. Because newer IT equipment can function at greater temperatures securely than older ones, the cloud infrastructure may not need to be as cold as it formerly was.
- **Power Management**—After many minutes of inactivity, these functions may be configured to automatically conserve power hard drives and monitors. Increased interest in generating cost-effective energy for new enterprises has been fueled by rising power rates. Once computers are dumped in landfills, toxic chemicals and unsafe quantities of lead and flame retardants are discharged. In general, green computing is the study of making computer use more ecologically friendly. The purpose of this paper [3] is to investigate several methods for reducing energy consumption in an operating system without sacrificing system performance parameters.
- **Disposing e-waste**—The need for information technology (IT) in many aspects of life, as well as automation in numerous industries, has boosted the demand for various computer devices dramatically. As a result, the quantity of energy and power required by these gadgets is increasing, raising the amount of carbon of the environment and the energy expenses of the global IT sector. The alarming rise in carbon dioxide levels in the atmosphere has prompted the development of more

environmentally efficient and environmentally green computing. This chapter [4] addresses how to reduce e-waste by properly recycling electronic components. e-waste should be disposed of in accordance with federal, state, and municipal rules.
- **Using alternative energy sources**—Alternative energy sources, including such geothermal cooling, wind, and hydroelectric power, should be investigated.
- **Upgrading the data centres**—Older equipment often consumes more energy and generates more heat than modern equipment. Hot and cold aisle configurations may be utilized to arrange assets depending on their energy usage and temperature, maximizing the efficiency of heating, ventilation, and air conditioning.

IoT is a new idea that intends to link billions of objects. The IoT devices perceive, gather, and send data from their environments. This huge flow of data among billions of devices requires a vast quantity of energy. Green IoT is a concept that aims to reduce the power consumption of IoT devices while also protecting the environment. The authors [5] began by providing an overview of green IoT and the issues that are encountered as a result of the excessive use of energy-hungry IoT devices, which is motivated by developing a sustainable environment for IoT. Green computing aids in the reduction of energy use, resulting in decreased CO_2 emissions. Because of a decrease in the usage of fossil fuel in power stations and mobility.

Green computing is one of the most effective methods for using resources like as heat, light, and electricity in a sustainable manner. Because Green Computing allows you to make more effective use of resources like power and office space. The operating expenses are significantly reduced. Green computing promotes enterprises and government agencies to recover e-waste and optimize energy use in order to have no or little environmental effect. Green Computing allows businesses achieve compliance and regulatory requirements, which helps them, enhance their corporate image. The Green Internet of Things aims to improve the efficiency in IoT concepts, and the quicker we handle this issue, the better. Green IoT is described as energy-efficient solutions in IoT to either lessen or eliminate the green-house impact generated by current applications or in IoT itself. In the first possibility, the Internet of Things will aid in the elimination of the greenhouse effect; however, in the second instance, the Internet of Things will be further improved to prevent the greenhouse effect.

2 Exposing Applications of IoT in Green Computing

Hardware and software aspects [6] for green internet technologies should be considered, with hardware solutions producing gadgets that use less energy without sacrificing performance. Software products [7], on the other side, provide modern systems that use the least amount of resources while using less energy. Virtual machine power-saving measures should also be incorporated. As highlighted in Fig. 2, there are many applications of IoT in green computing.

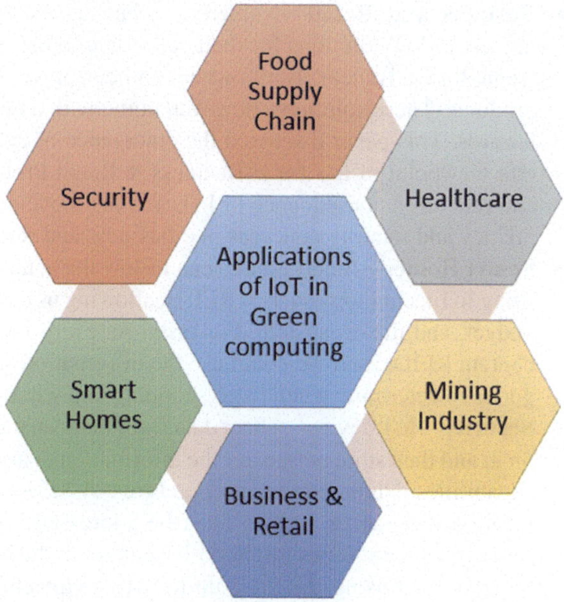

Fig. 2 Applications of IoT in Green computing

- **Food Supply Chain**—IoT has the potential to transform the commercial world. Vendors may trace their goods' manufacturing from the farm to the end customer using IoT technology. This study [8] offers a business-oriented IoT model for the food supply chain that may improve food security and used to gather data on manufacturing processes, which can then be controlled to make better business choices.
- **Healthcare**—People's lives are being affected by the introduction of IoT in the healthcare sphere. The use of cutting-edge wearable technologies is making life easier for those who are afflicted with a sickness. These wearable gadgets are designed to have certain qualities that make them accepted as a part of the human body. Edge computing and other cutting-edge technologies are supporting wearables in delivering the required performance. This study [9] aims to emphasize the relevance of wearables in IoT healthcare, as well as the reference architecture of the current wearable technologies and different IoT communication protocols. In addition, the features of the various wearables employed in IoT healthcare are highlighted in this study. Because IoT-based healthcare apps vary from those that operate on conventional wireless networks, the demand for security is significantly greater. As a result [10], smart portals are competent of dealing with a wide range of difficulties that ubiquitous healthcare systems encounter, including as dependability, adaptability, and security.
- **Mining Industry**—IoT technology may be utilized to protect miner safety and offer mining companies with vital information about the mining process, allowing them to improve their present methods. To increase interaction between miners and their employers, RFIDs, WiFi, and sensors may be used.

- **Business and Retail**—Numerous e-businesses and retail establishments are turning to IoT solutions for their sales, marketing, efficiency, and advertising as technology advances. Both owners and customers benefit from these IoT-based products. These solutions, however, come with a number of security and privacy hazards. This paper discussed the emergence of cyber threats in IoT, the corporate viewpoint of IoT for E-Business & Retail Protection, changes in e-business and retail as a consequence of IoT, and the movement of threat agents related to privacy and security concerns in e-business and retail [11].
- **Smart Homes**—Smart houses can follow the actions of residents in the building using IoT technology such as RFIDs, allowing us to make choices that save power, budget, and the ecosystem. For instance, every item within a smart fridge may contain RFIDs, and we could use the information supplied by the sensors on the goods to determine when to go shopping and what we need to purchase.
- **Security**—IoT devices collect data from their surroundings, conduct calculations on it, and then store or transfer the information to another device with comparable capabilities. Numerous security and trust risks occur as a result of this communication mechanism, which must be addressed immediately. Certain trust and security vulnerabilities might pose a danger to the IoT ecosystem, thus they must be addressed using the most up-to-date architecture, methodology, and equipment. This article [12] aims to illustrate some of the most significant risks to IoT security and trust, as well as how current design addresses them.

To fulfil the ever-increasing need for IoT, cloud—based services are deployed. Data centres are quickly becoming one of the top energy users in order to support the IoT paradigm's architecture. As more advancements [13] arise and technology adopts new habits, the need for energy will rise in the future, culminating in the adoption of green computing. The use of two popular and emerging technologies, Cloud Technology and the Internet of Things, is now a hot topic in the fields of agriculture and healthcare. This article [14] examines a variety of technologies and concerns related to green cloud technology and green Internet of Things, as well as the decrease in power consumption of the two approaches when used together in agriculture and healthcare.

3 IoT in Green Computing—Challenges and Issues

A variety of techniques need be followed in order to enforce the Green IoT, and many existing technologies for Green IoT are offered in. A framework for the efficient energy optimization of IoT items was suggested in order to deploy Green IoT. Green IoT may also be deployed with the help of green RFIDs, green datacenters, green sensor networks, and green cloud computing. IoT is a new technology that is altering our perceptions of the IT business. IoT will have a significant influence on how we approach certain difficulties in our everyday lives, and it will undoubtedly make our lives simpler and better, but with ease comes complications. We must cope with large-scale energy resource use, and the IoT will be more efficient in this regard.

IoT [15] is capable of linking small-scale to large-scale technologies. Such billions of gadgets constitute a vast network of devices on a massive scale that we refer to as the Internet of Things. Despite recent technological advancements, there has been a significant spike in the amount of IoT devices, which has resulted in an increase in energy consumption. This sparked interest in low-power IoT devices. There are several uses and advantages of employing IoT enabled devices; yet, studies have revealed that there is a possibility for a significant number and severity of important environmental concerns, such as excessive energy usage, carbon emissions, and other industrial effluent.

To address the negative consequences of such widespread use of ICT devices, the scientific community has intensified its efforts in the areas of Green IoT and Networking. Luckily, technologies such as the IoT may be combined with green technology to decrease electrical waste and the usage of comparatively less power for potential developments, resulting in better outcomes for our ecosystem. This chapter [16] examines the relevance of green computing and IoT for smart cities, as well as real-world practical implications. The green IoT is a new topic that has piqued the interest of academics and businesses since it offers energy-efficient services and allows for the generation and use of sustainable power. Nevertheless, blockchain technology [17] has emerged as a widely used IoT technology, receiving significant attention from energy corporations, start-ups, banking institutions, authorities, and academics. This research focuses on the role of developing blockchain technology in the green IoT ecosystem, offers the critical aspects that must be considered in order to establish a green IoT ecosystem, and examines how blockchain technology helps to green the IoT ecosystem. It also discusses outstanding concerns and research directions paths to address when developing a green IoT ecosystem utilizing blockchain technology. With an emergence to blockchain technology, its uses, and problems, the chapter [18] will address the function of blockchain in digital forensics. It also delves into the blockchain technology's structure and procedures. The chapter also discusses how to manage digital evidence by using Ethereum and Hyperledger to preserve the chain of evidence.

There are several challenges to the IoT and green computing, some prominent of them are mentioned below.

- Green IoT infrastructure is becoming less complex.
- The trade-off between effective dynamic bandwidth sensing and effective spectrum sensing.
- To have the minimum impact on the environment, applications should be green.
- Power consumption models and the dependability of green IoT
- Efficient energy mechanisms for IoT, such as wind, photovoltaic, vibration, and thermal, are necessary to make IoT viable.
- Security mechanisms that are effective, such as encryption and controlling devices
- Energy-efficient IoT solution with context-awareness.
- Energy-efficient communication equipment and standards should be utilized.
- Power-saving cloud management (Table 1).

Table 1 Comparative analysis of literature in IoT and Green computing

Author	Year	A	B	C	D	Major findings
Liu et al. [19]	2019	✓	✓	✓	✓	The authors advocated that IoT be greened in 3 components: ambient green energy gathering, green energy wireless charging, and green energy balancing, with the latter step reinforcing the first to assure green energy provision
Ferrag et al. [20]	2020	✓	✓	✓	✓	This article discusses research obstacles in the realm of green IoT-based agriculture, including security and privacy concerns. We begin by outlining a four-tier green IoT-based farm architecture and reviewing current smart agriculture surveys. The threat scenarios against green IoT-based agriculture are then divided into different categories, including assaults on privacy, identification, confidentiality, availability, and integrity features
Borkar et al. [21]	2021	✓	✓	✗	✓	The majority of transport vehicles are Wi-Fi connected, which necessitates additional energy usage. Certain effective methods should be developed in order to reduce increased energy use. This chapter introduces the Green IoT, discusses its problems and concerns, the different technologies utilized in Green IoT, and the necessity for Green IoT in intelligent transportation, as well as a case study
Ghorpade et al. [22]	2021	✓	✓	✗	✗	Massive expansion in IoT-enabled wireless sensor node installation is projected in the near future, thanks to the rise of the Internet of Things and machine-to-machine interactions. Energy efficiency, optimal route computation, and total transmission cost are three major problems for the sensor network. We present a new multi-objective optimisation agent due to the particle swarm grey wolf optimisation and reverse fuzzy ranking to minimize bias toward one of the goals and also to make position update easier

(continued)

Table 1 (continued)

Author	Year	A	B	C	D	Major findings
Mekala et al. [23]	2018	✓	✓	✗	✗	For X-IoT implementations, this study examines energy-efficient sensor and resource-based VM selection algorithms. It is demanded that measurement functions, topologies, and VM scheduling mechanisms be distinguished. Sensor energy usage, edge computing components, and communication techniques all limit the exact operational specifics of chosen publications

A: Green computing B: IoT C: security D: cloud computing

4 Conclusion

The Internet of Things (IoT) is becoming more popular throughout the world, with its network architecture consisting of numerous devices such as actuators, sensors, and other devices that are physically implanted items capable of perceiving, processing, and communicating information over the network. Physical items are effortlessly incorporated into the information network, allowing them to become active members in business operations, thanks to the Internet of Things. As a result, information on their state, the surroundings, manufacturing, and scheduled maintenance with planning, and much more may be sent. In this chapter, the author have exposed the various important applications of IoT in Green computing and the challenges associated with them. Furthermore, the chapter also highlights the green computing tactics in the domain of IoT.

References

1. Shuja, J., et al.: Greening emerging IT technologies: techniques and practices. J. Internet Serv. Appl. **8**(1), 1–11 (2017). https://doi.org/10.1186/S13174-017-0060-5/FIGURES/3
2. Poongodi, T., Ramya, S.R., Suresh, P., Balusamy, B.: Application of IoT in Green computing. Advances in Greener Energy Technologies, pp. 295–323 (2020). https://doi.org/10.1007/978-981-15-4246-6_19
3. Dar, S.K., Asif, S., Islam, A.M.: Power management and green computing: an operating system prospective. https://www.researchgate.net/publication/338215315_POWER_MANAGEMENT_AND_GREEN_COMPUTING_AN_OPERATING_SYSTEM_PROSPECTIVE. Accessed 06 Feb 2022
4. Roy, S.: Green computing : an eco-friendly and energy-efficient computing to minimize e-waste generation. Go Green for Environmental Sustainability, pp. 87–100 (2021). https://doi.org/10.1201/9781003055020-07

5. Karmankar, P.S., Tadse, P.P., Karmankar, P.S., Tadse, P.P.: Application of IoT in Green computing. JETIR **8**(1), 111–114 (2021). Accessed 06 Feb 2022 [Online]. https://www.jetir.org/view?paper=JETIREK06028.
6. Garg, A., Singh, A.K.: Applications of internet of things (IoT) in Green computing. Intelligence of Things: AI-IoT Based Critical-Applications and Innovations, pp. 1–34 (2021). https://doi.org/10.1007/978-3-030-82800-4_1
7. Albreem, M.A.M., et al.: Green Internet of Things (IoT): an overview
8. Yadav, S., Luthra, S., Garg, D.: Internet of things (IoT) based coordination system in Agri-food supply chain: development of an efficient framework using DEMATEL-ISM. Oper. Manag. Res. **2020**, 1–27 (2020). https://doi.org/10.1007/S12063-020-00164-X
9. Singh, K., Kaushik, K., Ahatsham, Shahare, V.: Role and impact of wearables in IoT healthcare. Adv. Intell. Syst. Comput. **1090**, 735–742 (2020). https://doi.org/10.1007/978-981-15-1480-7_67
10. Kaushik, K., Dahiya, S., Sharma, R.: Internet of things advancements in healthcare. Internet of Things, pp. 19–32 (2021). https://doi.org/10.1201/9781003140443-2.
11. Kaushik, K., Dahiya, S.: Security and privacy in iot based e-business and retail. In: Proceedings of 2018 International Conference on System Modeling & Advancement in Research Trends (SMART), SMART 2018, pp. 78–81 (2018). https://doi.org/10.1109/SYSMART.2018.8746961
12. Kaushik, K., Singh, K.: Security and trust in IoT communications: role and impact. Adv. Intell. Syst. Comput. **989**, 791–798 (2020). https://doi.org/10.1007/978-981-13-8618-3_81
13. Muniswamaiah, M., Agerwala, T., Tappert, C.C.: Green computing for internet of things. In: Proceeding of 2020 7th IEEE IEEE International Conference on Cyber Security and Cloud Computing, 2020 6th IEEE International Conference on Edge Computing and Scalable Cloud, CSCloud-EdgeCom 2020, pp. 182–185 (2020). https://doi.org/10.1109/CSCLOUD-EDGECOM49738.2020.00039
14. Nandyala, C.S., Kim, H.-K.: Green IoT agriculture and healthcare application (GAHA). Int. J. Smart Home **10**(4), 289–300 (2016). https://doi.org/10.14257/ijsh.2016.10.4.26
15. Ahmad, R., Asim, M.A., Khan, S.Z., Singh, B.: Green IoT—issues and challenges. SSRN Electron. J. (2019). https://doi.org/10.2139/SSRN.3350317
16. Sarkar, N.I., Gul, S.: Green computing and internet of things for smart cities: technologies, challenges, and implementation. Green Energy and Technology, pp. 35–50 (2021). https://doi.org/10.1007/978-3-030-48141-4_3
17. Sharma, P.K., Kumar, N., Park, J.H.: Blockchain technology toward green IoT: opportunities and challenges. IEEE Netw. **34**(4), 263–269 (2020). https://doi.org/10.1109/MNET.001.1900526
18. Kaushik, K., Dahiya, S., Sharma, R.: Role of blockchain technology in digital forensics. Blockchain Technology, pp. 235–246 (2022). https://doi.org/10.1201/9781003138082-14
19. Liu, X., Ansari, N.: Toward Green IoT: energy solutions and key challenges. IEEE Commun. Mag. **57**(3), 104–110 (2019). https://doi.org/10.1109/MCOM.2019.1800175
20. Ferrag, M.A., Shu, L., Yang, X., Derhab, A., Maglaras, L.: Security and privacy for Green IoT-based agriculture: review, blockchain solutions, and challenges. IEEE Access **8**, 32031–32053 (2020). https://doi.org/10.1109/ACCESS.2020.2973178
21. Borkar, P., Balpande, V., Aher, U., Raut, R., Borkar, M.S.: Green IoT for smart transportation: challenges, issues, and case study. Green Internet of Things and Machine Learning: Towards a Smart Sustainable World, pp. 269–296 (2021). https://doi.org/10.1002/9781119793144.CH10
22. Ghorpade, S., Zennaro, M., Chaudhari, B.S.: Towards green computing: intelligent bio-inspired agent for IoT-enabled wireless sensor networks. Int. J. Sens. Netw. **35**(2), 121–131 (2021). https://doi.org/10.1504/IJSNET.2021.113632
23. Mekala, M.S., Viswanathan, P.: A survey: smart agriculture IoT with cloud computing. In: 2017 International Conference on Microelectronic Devices, Circuits and Systems (ICMDCS), vol. 2017-January, pp. 1–7 (2017). https://doi.org/10.1109/ICMDCS.2017.8211551

IoT Safe Area: An Innovative Approach for a Smart Decentralized and Hybrid Decision Support Process for Industrial Warehousing

Trab Sourour, Zouinkhi Ahmed, Bajic Eddy, and Abdelkrim Mohamed Naceur

Abstract This chapter shed lights on a decentralized approach which aims at monitoring resources securely and dealing with products in IoT-based warehouses, and this is part of hybrid logistic decision-making aided processes. These latter are composed of two phases. The first phase consists in logistic centralized decision-making in relation to terminal product allocations, utilizing advanced product classifications, floating locations and the method of static potential fields. The second phase consists in decentralized decision making, hence assuring at the shop floor level the reactive product-trajectory control through the use of the method of dynamic potential fields, while taking into account product-resource interactions. The proposal can achieve that safer warehouse management which is based on the reactivity of smart shelves and products against any critical surrounding risk. Some scenarios for negotiated interaction mechanisms that concern the safe storage and picking operations are proposed. Using intelligence in Warehouse Management System (WMS) results in proposing an "IoT-controlled Safe Area" (IoT-SA) as a new concept. Our contribution is to make decisional, communicational and informational abilities near the warehouse because of those communicating objects. Thus, we can achieve safety assurance while decreasing certain delays in decision-making and increasing the effectiveness of solving dynamic and local disruptions. This leads to avoid the inherent WMS-centralization short comings. Therefore, we present an implementation case for IoT-SA concept, which utilizes Labview software and a platform of a ZigBee wireless sensor network.

T. Sourour (✉) · Z. Ahmed · A. Mohamed Naceur
MACS Laboratory: Modeling, Analysis and Control of Systems, National Engineering School of Gabes, Gabes, Tunisia
e-mail: sourour.trab@gmail.com

Z. Ahmed
e-mail: ahmed.zouinkhi@enig.rnu.tn

A. Mohamed Naceur
e-mail: naceur.abdelkrim@enig.rnu.tn

B. Eddy
CRAN-CNRS UMR7039, Université de Lorraine, Nancy, France
e-mail: eddy.bajic@univ-lorraine.fr

S. El Himer et al. (eds.), *Artificial Intelligence of Things for Smart Green Energy Management*, Studies in Systems, Decision and Control 446,
https://doi.org/10.1007/978-3-031-04851-7_11

Keywords Hybrid decision-aided approach · IoT-SA · WMS · Safety · Communicating object · Interaction mechanisms

1 Introduction

Cyber-Physical Production Systems (CPPSs), which rely in the newest developments in any information and communication technology as well as in manufacturing sciences and technologies, is the key shift of the fourth Industrial Revolution, which is known as Industry 4.0 [1]. Leveraging cutting edge technologies like Internet of Things (IoT) and machine-to-machine communication enables objects to be controlled or sensed in a remote way across the existing network infrastructure and creates a good opportunity to directly integrate the physical word into the cyberspace [2]. We can embed logistical IoT applications in railways, industries and airports, using GIS, RFID and a Wireless Sensor Network Platform (WSN) [3]. Integrating the IoT and CPPSs into logistics permits tracking assets in real-time, handling transport in a better way and managing risks accurately. According to [4], Industry 4.0 perspectives may be real if logistics provides the needed input factors for production systems in the right place, with the right quality and at the right time. Within warehouse logistics, while knowing that logistic, safety and sustainability problems have been important challenges of modern industrial enterprises, the IoT is that relevant concept for guaranteeing efficient management and safety performance.

Accidents in warehouses have become common due to many factors. A recent research study of 78 Dutch warehouses illustrated that an increase in "safety consciousness" would improve the individual consciousness of safety matters in order to avoid hazardous behavior [5]. The highest rate of dangerous accidents was strongly linked to transportation steps in marshaling yards and port areas and to intermediate temporary storage [6].

It is crystal clear that warehouse design management and control are very important for the supply chain efficiency [7]. The main topic about decreasing logistics costs and boosting safety level should be guaranteed by effectively optimized [8] and reactive warehouse management [9]. The main considered aspects are the exact knowledge of the inventory level, the optimum storage locations of the products and their states, as well as the safety monitoring of the interaction between products, resources and workers [7]. Newly introduced solutions emphasizing these goals could better improve the availability, efficiency, and safety in warehouse management.

The current chapter suggests a decision-making-aided hybrid logistic approach to safely monitor and handle products and resources in IoT-based warehouses. We can transform these products into smart agents capable of deciding any storage destination. Such a warehouse can be built according to centralized decision-making stage based on logistics as well as on the safety product data. This can also be combined with a decentralized and reactive stage to permit us to monitor the state of products while moving toward the best storage location.

This chapter comprises 4 sections. The first one is dedicated to the topicality of industry 4.0 in logistics and safety matters in WMS. The second section briefly highlights logistic issues in WMS and related work in the CPPS research field. The third section is devoted to our proposed model for a hybrid decision making and control for WMS operation management. In the last one, an industrial case implementation is discussed. Interaction mechanisms in IoT-SA are used in a WSN while deploying the Waspmote technology, which allows identifying risks according to the Radio Signal Strength Indicator (RSSI) measure and ensuring distance estimation in IoT-SA. The IoT-SA control of interactions interface is presented and developed with Labview Supervisory Control and Data Acquisition (SCADA) software. Future developments are discussed and concluded at the end of the paper.

2 Related Works on Logistic Issues in WMS and CPPS

A WMS handles products in multiple operations like shipping, picking, storing and receiving, which implies the interaction of products with operators and machines. In fact, they all have at each product life cycle a specific impact [8]. The authors in [9–11] highlighted those major keys with the aim of ensuring that safety issues within decision making models would be precise, including effective gathering and monitoring as well as managing information using hybrid, distributed or centralized control approaches. In addition, some core functions are in particular significant for efficient and optimized warehouse management such as product allocation, picking and identification.

2.1 Warehousing Operations

The first task of product allocation is a serious problem because of its impact on the performance as regards the costs of operations and the level of service quality in warehouses [12, 13]. This task consists of the allocation of products to various slots of shelves. At the same time, space occupation is increased and handling costs are decreased. In addition, a product class-based storage technique was put forward in [10] based on compatibility constraints and a mathematical approach, which allowed partitioning product items.

Secondly, the most costly and labor-intensive activity is the picking operation which represents about 55% of the warehouse operating costs [5]. Put it simply, the picking delay cost consists in general in the sum of delay costs which occur when executing a picking plan and at the same time preparing a product package to be shipped.

To decrease the picking penalty cost, a multi-agent architecture was put forward by [7] utilizing a distributed IoT infrastructure for controlling and monitoring virtual agents linked with objects in the physical warehouse. Despite their vital roles in the operations at the warehouse, the human operation and warehousing resource interactions as regards safety were ignored.

Besides, a week management of warehouse spaces will affect the staff cost and results in inappropriate utilization of shelving arrangement. As a matter of fact, shelves can be considered full while they are empty. During warehousing routine, workplace safety inspections will be carried out by human operations, which can make the risk level higher.

Eventually, 'real-time' product information availability and product identification are notable problems. Using RFID technology is a key shift solution for speeding up information flows and making them more visible [14].

Our contribution is to reinforce decentralized interaction capabilities in a warehouse, using proactive negotiation mechanisms between materials and humans, in order to overcome safety related weaknesses and limitations.

2.2 *Safety and Product-Centered Approach*

Products collaboration is a relevant research topic in academic and industrial activities [15]. In [16] a framework for collaborative products networking was proposed with specific implementations in an electric water heater. In earlier work, a new Stigmergy-concept interpretation [17] was proposed, in which they achieved a good cooperation between production actors and that was due to specific attributes named informational pheromones, which were carried out by RFID-tagged products.

An Intelligent Product (IP) paradigm [18, 19] was established as a step beyond the RFID-tagged product use, which resulted in a new approach of product-driven manufacturing and logistics [20]. This IP can disseminate information when traveling via communication with its neighbors and participating in an active way in its future.

Although very elaborated, those research studies do not cover, or take into consideration, safety issues. Thus, although safety in warehouses has become very important and sustainable development and environmental protection problems have emerged, we still have insufficient research in relation to the safety issues in logistics [21]. Commonly, operator safety has the threat of manually handling material during non-routine operations, which we may improve by methods and techniques which allow us to control and monitor the interaction between materials and operators [22].

Since then, the IoT concept [23] has become a promising strengthening tool for safety management of product warehousing thanks to enhanced remote dynamic supervision capability, as demonstrated in [11]. Also, in [23], IoT was integrated with cloud computing, through the use of RFID and Internet technologies, with the objective of automatically identifying products and avoiding accidents.

The study of related works shows a clear change in tackling logistics issues from the process centric activities to the product centric approach. The product tends to become informant, communicating and smart enough to take part into the decision process. Nevertheless, current work still lacks consideration of the security of goods and people, as well as safety control and low-level interaction between products and humans at the shop floor level. Facing the rapid development of automatic decision making in WMS and at a larger scale in Industry 4.0 context, these matters are getting higher importance for the full achievement of future smart production and logistic systems.

Our proposal intends to contribute to these achievements by mean of a conceptual and practical solution by providing a hybrid decision making-aided model, based on an IoT approach, for logistics and warehousing.

3 Hybrid Decision Making-Aided Model for WMS

We propose in this research work a hybrid decision making-aided model (see Fig. 1): a hybrid first phase for the "centralized logistics decision" approach (steps 1, 2, 3, 4 in Fig. 1) and a second phase for "decentralized control" (steps 5, 6, 7 in Fig. 1) on the shop floor [24].

3.1 Logistic Decision Making: Centralized Phase

Various data are provided by the WMS as regards decision-making with a central point of view with the purpose of mastering and controlling products [24]:

- Difficulty of grasping, product rotation frequency, product weight, … These have been utilized for calculating the allocation cost of every product, and they are called **Logistic Data (LD)**.
- Environmental constraints of storage locations and constraints between the products allowing the classification of products into homogeneous classes, and they are called **Product Constraints (PCs)**.
- Available storage locations, **Floating Location Emplacement** $FLE_{m,n,p}$, which will be determined through the use of a neighbors' security test ($Safety\_level$ assignation for each location based on their compatibility degree with products). All remainder locations and **occupied locations** will be considered as not allowed to store products, to which a $Safety\_level_{prohib\_loc}$ is assigned.

Classifying products is the first step of the proposed approach (step 1 in Fig. 1). The second step is the product allocation cost calculation (step 2). Thirdly, there is the distinction between suitable $FLE_{m,n,p}$ storage locations of a product from prohibited ones (step 3). Fourthly, a safety level for prohibited locations, in particular, is assigned (step 4). Finally, the calculation step of the resulting potential field (repelling/attraction) is executed (Step 5 in Fig. 1). This step is based on the potential field theory allowing deciding in the warehouse a secure and more optimal storage product destination. The assignment of product allocation costs allows choosing the most appropriate destination to locate products (optimizing allocation costs). Products having a high cost are allocated close to the warehouse exit door.

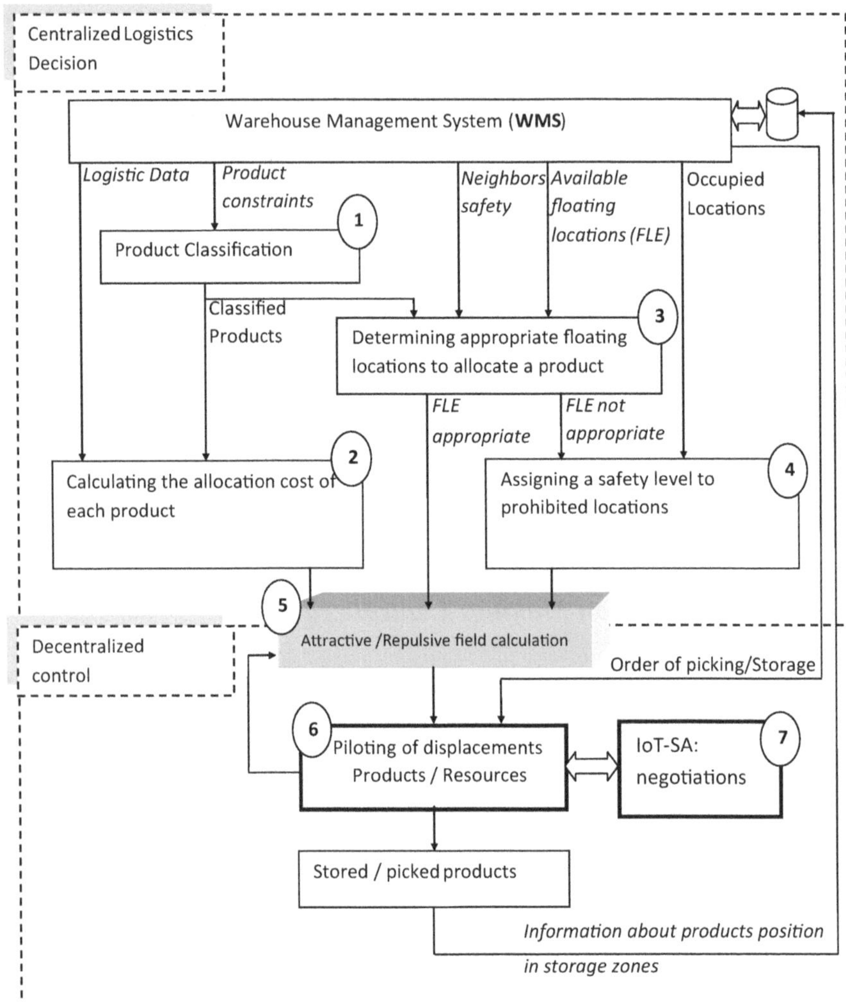

Fig. 1 Overall block diagram for principle of decision making-aided approach

3.1.1 Products Classification Method

The aim of product classification is to reduce the risk factors of accidents occurring while controlling the product movements. Passing from processing based on "products" another based on "product classes" based treatment allows to optimize allocation functions. A methodology that aimed for classifying a set of constrained products was proposed utilizing the graph-coloring theory.

The product-grouping process begins with the creation of a matrix of binary constraints between products. This matrix will be transformed to a constraint colored graph algorithm DSATUR [25] (degree of saturation) well known for coloring graphs with COSINE algorithms. DSATUR is a very effective method for coloring graphs [26].

The $(Safety\_level)$ parameter is allocated to all product classes with the target of setting up priority-based storage determined by the urgency against accidents.

3.1.2 Calculation of Product Allocation Cost

After the classification of products comes the phase of allocating them which needs to consider factors related to the nature and reciprocal constraints between products which will are called **Product Safety Data** (PSD) including the dangerousness of products, and **Logistic Data** (LD) which characterize the product physics and logistic flow characteristics. We consider three main characteristics among LD that influence the decision of its allocation to the most appropriate storage location:

- **Product weight** affects the choice of an appropriate storage location. It is quantified in two levels (low and high) depending on product mass. Table 1 illustrates the correspondence between the real weight of a product and the weight cost function. N is determined according to the available means of transport in the warehouse, the suitability of the manipulators, etc.
- **Difficulty level of product grasping** characterizes the difficulties related to the product regarding the specific resources of transport (conventional craft, forklift, manipulator arm…). It is quantified in three levels (low, medium, and high) (see Table 1).
- **Frequency of product rotation** refers to the frequency of the product entry and exit from stock. This characteristic will take three levels (low, medium, and high) (see Table 1).

Table 1 Logistic data characteristics

	Low	Average	High
Product weight (M)	(M < N kg)		(M > N kg)
	Product weight cost: $K_M \in [17,60]$		$K_M \in]60,100]$
Grasping difficulty (D)	Manual storage	Can be stored using an average cost transport resource (conventional gear)	Requires special transport mean
	Grasping difficulty Cost; $D_F \in [17, 30]$	$D_F \in]30,60]$	$D_F \in]60,100]$
Product rotation frequency (F)	Product usable a few times/year	Product usable a few times/week	Product usable every day
	Rotation frequency cost: $F_R \in [17, 30]$	$F_R \in]30,60]$	$F_R \in]60,100]$

The LDs have been utilized for calculating the Allocation Cost of every product individually after assignment to a class Alloc_cost(P_{ij}). The allocation cost has an impact on the terminal product location (step 2 in Fig. 1). A product with the highest allocation cost must be positioned closer to the exit door, and a lower cost can be allocated at positions further away from the exit door.

The allocation cost summarizes the product weight cost K_M, the grasping difficulty cost D_F and the rotation frequency cost F_R:

$$\text{Alloc\_cost}(\text{P}_{ij}) = \sum K_M(\text{P}_{ij}) + D_F(\text{P}_{ij}) + F_R(\text{P}_{ij}) \quad (1)$$

The allocation cost will be used after determining appropriate product storage locations as a factor influencing the calculation of attractive locations while driving product/resource movement in the warehouse (step 5 in Fig. 1).

3.1.3 Determination of Appropriate Floating Locations and Safety Level Assignment

First, we need to determine the appropriate Floating Location Emplacements $FLE_{m,n,p}$ based on a safety control approach (step 2 in Fig. 1). The aim is to choose the suitable location for a product $P_{ij} \in C_j$, depending on three parameters:

- Product characteristics (weight, nature: liquid/solid, …), the size of the location compared to the size of the product to be stored.
- Influence of the storage environment (temperature, humidity…) on product P_{ij}.
- Product compatibility with products already stored in the vicinity of $FLE_{m,n,p}$.

Then the remaining floating locations $FLE_{m,n,p} \backslash FLE_{m,n,p}(P_{ij})$ are considered prohibited for the allocation of this product. Moreover, a safety level ($Safety\_level_{prohib\_loc}$) must be assigned to all prohibited locations. Safety levels previously assigned are used as a repelling factor when applying the theory of potential fields and vary depending on the dangerousness of the targeted zone for product storage (Fig. 2).

3.1.4 Potential-Field Determination Steps

In many domains, the theory of potential fields has been proposed with different forms of mathematical calculations of attractive/repulsive fields such as the parabolic potential field [27] for robotic path planning, the logarithmic potential field [28] for the allocation of dynamic tasks and the exponential potential field for routing products in flexible production systems [29].

We take advantage of the theory of potential fields, after the execution of the product classification and calculating the costs of allocating products:

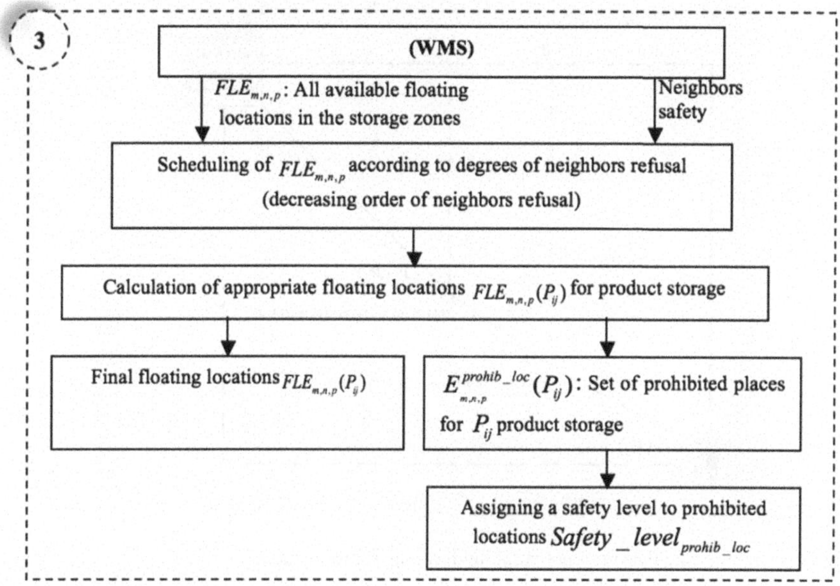

Fig. 2 Determination of appropriate floating locations for product storage

- Statically determining the best appropriate terminal location for products, as a function of LD and PSD criteria.
- Optimizing path of products, which allows reaching its terminal location. In fact, the two attractive and repulsive potential fields will be basically calculated for each product along its path and after that into its final destination through the use of a product direction-finding algorithm and according to the potential-field theory. The calculation will be done in a dynamic way.

In a first step, U_{al} (attractive potential field) is given to the available storage locations $FLE_{m,n,p}(P_{ij})$ of a P_{ij} product. This potential field depends on Alloc_cost(P_{ij}) (the allocation cost function) as a function of the LD and the distance between the current product position and its destination.

In a second step, a repulsive potential field U_{rk} is allocated to each product's prohibited locations $prohib\_loc$, which have different repulsion levels depending on their $(Safety\_level_{prohib\_loc})$ regarding the product to be stored.

$V_p(P_{ij})$ (resulting potential field) adds the attractive and repulsive fields and represents the best and safe attractive direction towards the terminal location.

$$V_p(P_{ij}) = \sum_{l=1}^{A} U_{al} + \sum_{k=1}^{B} U_{rk} \tag{2}$$

A: is the number of $FLE_{m,n,p}(P_{ij})$ appropriate locations, and B is the number of $prohib\_loc$.

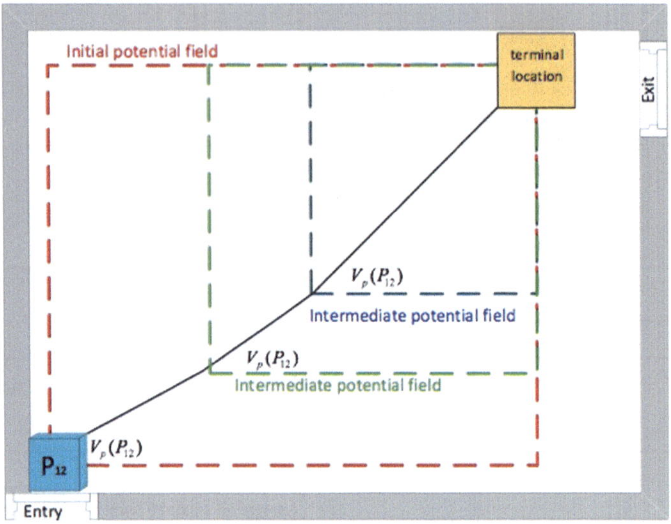

Fig. 3 Dynamic calculation of potential fields along P_{ij} movement

3.2 Decentralized Warehousing Control

The resulting potential field is recalculated in a dynamic way to optimize the product path and at the same time moving towards its terminal location. In case locations of destination storage are chosen, products will discover their optimal allocation paths by recalculating the resulting potential field $V_p(P_{ij})$ at each step movement (Fig. 3). Simultaneously, a negotiation model based on negotiations between products, resources and operators, linked with a smart IoT-Safety Area new concept will be a key shift of the decentralized approach which is developed and proposed in [30].

After finishing the movement phase, the data about the product position in the storing area will be uploaded to the centralized WMS.

4 Smart IoT-Controlled Safe Area New Concept

We propose an "IoT-controlled Safe Area" (IoT-SA) concept as intelligence embedding of warehouse elements that are organized independently by interactions between communicating objects. In addition, they are dedicated to safety assurance. This concept can structure a warehouse as a group of areas (zones). These are safety controlled, but unfortunately some of them are free-constraint. In these zones there are vertical storage volumes (shelves) as well as horizontal storage surfaces (storage floor zones). Actually, we can monitor and control safety areas so as not to let risky interactions happen between actors that generally belong to the WMS operations. We can mention here transport resources, shelves, human operators, and products.

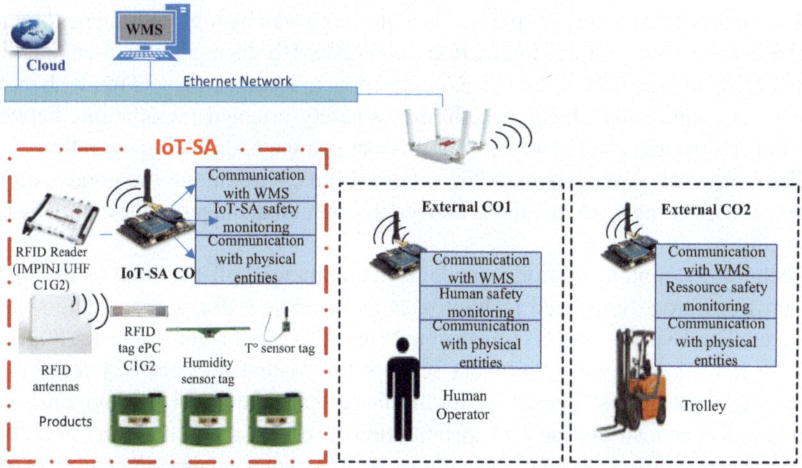

Fig. 4 Smart "IoT-SA" concept implementation

A three-layer shelf is presented in Fig. 4 as the center of the IoT-SA. This latter can be bounded by a 3D-envelope delimiting edges related to risky interaction zones between resident and incoming COs. A danger in the warehouse is present while picking goods for shipping preparations, when storing new orders or during their arrival, or while moving a product. To ensure total safety, some storage zones are turned to zones with safety-controlled interactions. Such transformed warehousing are hazardous and risky because of type and nature of products to be picked and placed, traffic density, environment safety level, etc.…

Criticism during operations is caused by permanent movements within close proximity to products, trolleys, and people. We can take the example of an operator carrying a product and at the same time using or not a trolley for storing products at appropriate shelf locations or when an operator picks some products from their shelves to prepare certain shipments.

The CO concept is instantiated upon every individual safety equipment (safety gloves, safety glasses, security helmet), trolley, shelf, human operator, and product. We equip each part with embedded communicating entities, so we attach the RFID tags to low-risk/cost entities or we embed in particular WSN nodes in human operators, products shelves, or trolleys. These entities are highly costly or risky. Product entities controlled by RFID tags have a virtual representation in the base station which involves the data needed to guarantee its intelligence and create its agents. Before giving the different negotiations and interactions between the warehouse COs in detail, firstly we define the physical meaning of every item.

A safe area is stable, at an initial situation, as regards safety risks. We compute the product placements in terms of their cross compatibilities and product data. Apart from any breakdown of products or change in ambient parameters, the area stability can only be disturbed actually due to the appearance of external entities (objects) such as operators, resources, or other products.

The "IoT-SA" concept is explained in a thorough way by highlighting negotiation mechanisms for most critical interaction cases related to storing products and picking tasks [30]. The intention of the IoT-SA concept is to leverage the CO model benefits of our communicating object, which allows safety-oriented negotiations between warehouse resources and actors during product storage and picking operations.

The industrial partner requires the safety of the environment, the goods and the people and avoiding work accidents and environmental damages to the company risk management policy.

Technology implementation is made from commercial off-the-shelf technologies chosen from broadly utilized technologies and standards for guaranteeing system sustainability. Every product is transformed into a CO by attaching a ZigBee 802.15.4 node to it while integrating ambient sensors for context awareness. On the other hand, the material cost impact limits the large-scale industrial deployment. As a consequence, it restricts the CO instantiation to expensive, dangerous or critical products. According to that, we can simply equip a non-critical product with a UHF-ePC RFID tag, which is assigned to a virtual product representation in the WMS base-station. Each tag includes characteristics related to the product type, its usage precaution and its impact on the environment.

Warehouse resources (trolleys, human operators) and IoT-SA are fully turned into the CO with Zigbee WSN nodes and ambient sensors. UHF RFID readers are mounted on each IoT-SA (i.e. shelves, storage spaces) with multiple antennas for covering the entire area with the target of ensuring product location and accessing to the data of products stored around. The IoT-SA CO oversees the IoT-SA management. In addition to that, it ensures the interaction with COs as well as with products situated in their decision-making space. It drives the safety policy by reacting when risks for goods or people security are detected and by blocking the threatening external COs.

In Fig. 5 the UML sequence diagram shows the way the IoT-SA CO interacts with locally stored RFID-tagged products and with external incoming COs. The sequence of messages illustrates the interaction with the COs in order to store and pick inside the IoT-SA. We automatically and spontaneously elaborate the negotiation scenario between the five involved main actors: the WMS base station, the RFID tagged products, the RFID reader with different antennas, the external CO (a human operator, a product, or personifying a trolley), and the IoT-SA CO. All new COs entering its space are checked by the IoT-SA CO and detected using an RSSI that is received from the COs (alt1). After that, we start a dynamic safety negotiation plan with the objective of controlling (accepting or rejecting) the entrance of the external COs within the IoT-SA (alt2) and a s a function of the operation type (picking alt3 or storing alt4) and according to the safety constraints. We represent a CO either an SE or a CE, which is an SE aggregation.

When the IoT-SA rejects an incoming CO, it will alert the centralized WMS and can stop the resource.

A safety-based WMS that implements the IoT-SA concept has been jointly experimented with the Tunisian Chemical Group in Gabes, Tunisia.

Actually, we discuss the industrial case implementation. The interaction mechanisms in IoT-SA are deployed in the wireless sensor network platform using the

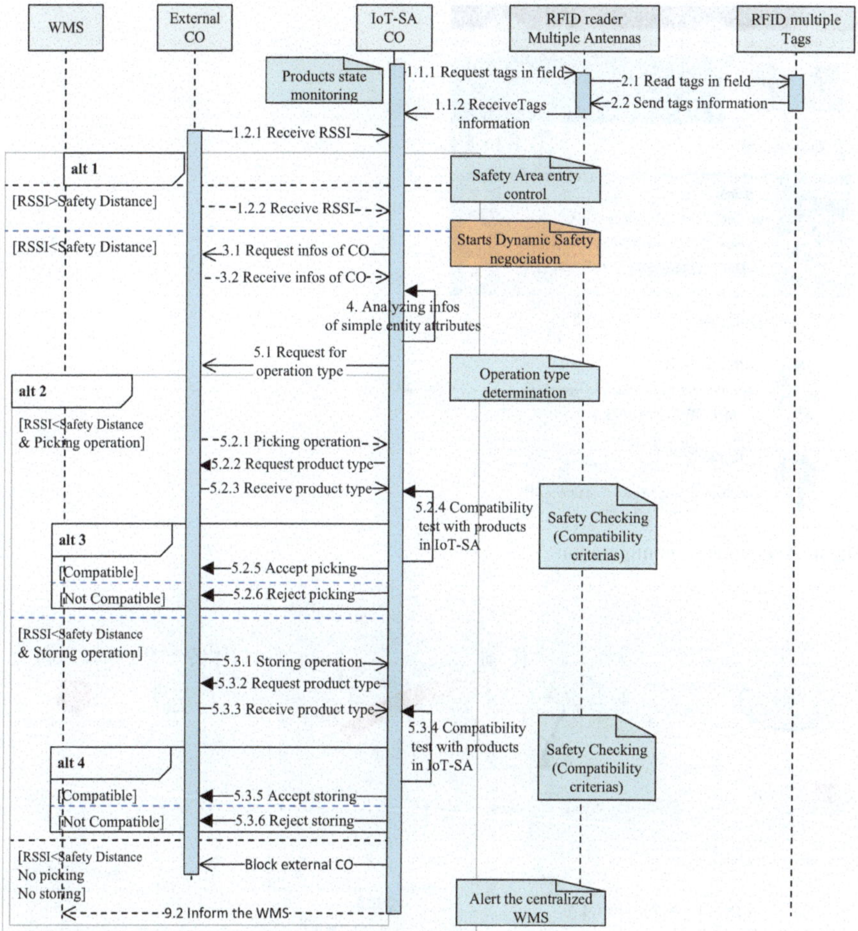

Fig. 5 UML sequence diagram of negotiation between IoT-SA actors'

Waspmote technology. Thus latter permits identifying risk according to the RSSI measure and ensuring distance estimation in IoT-SA. The IoT-SA control of the interaction interface is presented and developed with Labview SCADA software. IoT-SA CO and Product CO are configured using XCTU as routers/end devise API and the gateway attached to the laptop is configured as an API coordinator (Fig. 6).

Reject of IoT-SA entry (reject storing or reject picking): RSSI measurements between external CO "produit 1" and IoT-SA CO must be made until reaching the threshold fixed by distance tests (set at RSSI = 24 for entry to the IoT-SA zone). Once the external CO (attached to one new product) is in the zone, a compatibility test will be done and refused in case of incompatibility (Fig. 7).

Fig. 6 Xbee module configuration

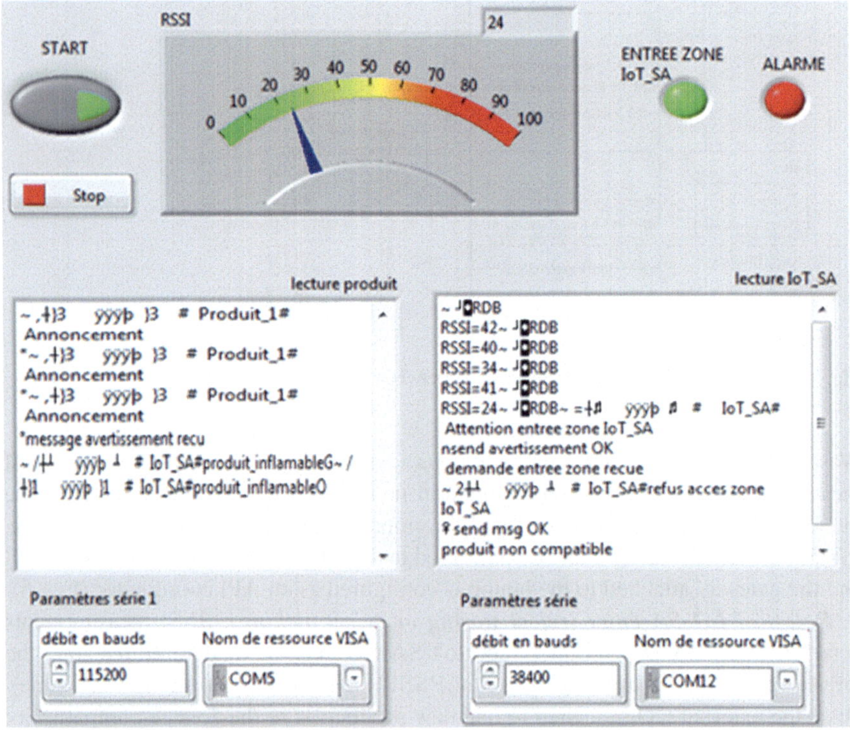

Fig. 7 Incompatibility between products and *reject of IoT-SA entry*

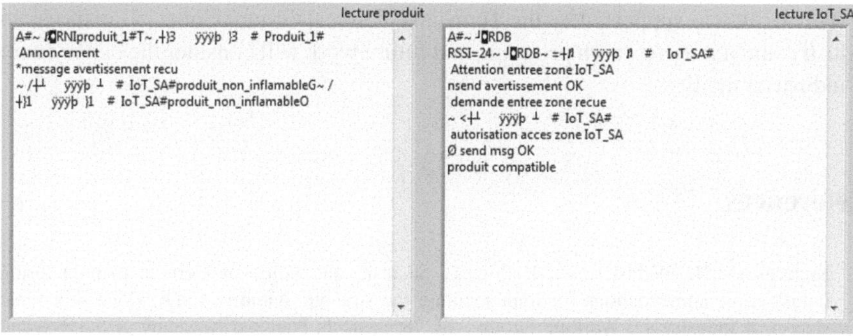

Fig. 8 Compatibility between products and *acceptance of IoT-SA entry*

Accept of IoT-SA entry (accept storing or picking): While entering in IoT-SA zone a compatibility test will be done between external CO "produit 1" and product in IoT-SA zone and accepted in case of compatibility between products (Fig. 8).

5 Conclusion

The deployment of the IoT concept offers new opportunities to achieve global and pervasive interoperability's between the environment, humans, machines and objects, as well as innovative relationships and services.

The presented work is oriented towards an IoT-based product-controlled system approach and takes part in the cyber physical production system paradigm, with an emphasis on safety issues in warehousing operations. Safety storage and transportation operations can be ensured by directly controlling it with devices and products. It can be applied also on persons acting jointly in warehousing in an ultimate way through monitored and controlled interactions of IoT entities.

The hybrid approach proposed by this chapter can achieve high-level safety in WMS controlling, through the reduction in the decision-making delay as well as ameliorating the solving effectiveness of dynamic disruptions. Such a technique mixes centralized and decentralized approach, aiming to avoid the inherent shortcomings of WMS which is commonly centralized and leveraging on decentralized adaptability. The hybrid decision-making aided model, in order to allocate products, uses Potential Field theory in centralized/distributed way and negotiation mechanisms in smart products based warehouse.

We validate the efficiency of our proposal by an interaction mechanism in IoT-SA deployment within a wireless sensor network platform using Waspmote technology, which allows identifying risks according to the RSSI measure and ensuring distance estimation in IoT-SA. The IoT-SA control of the interaction interface and compatibility test is presented and developed with Labview SCADA software.

This work was supported by the Tunisian Chemical Group (Gabes, Tunisia) in an industry–university research program, and future work will consider the deployment in industrial facility.

References

1. Kagermann, H., Wahlster, W., Helbig, J.: Securing the future of German manufacturing industry: recommendations for implementing the strategic initiative INDUSTRY 4.0. Final Report of Industry 4.0 Working Group, vol. 78. acatech, National Academy of Science and Engineering, München (2013)
2. Barkalov, A., Titarenko, L., Mazurkiewicz, M.: Introduction into embedded systems. In: Foundations of Embedded Systems. Studies in Systems, Decision and Control, vol. 195. Springer, Cham (2019). https://doi.org/10.1007/978-3-030-11961-4_1
3. Behnam, A., Sarkeshikian, A., Shafia, M.-A.: A study on IOT applications and technologies in logistics. In: Molamohamadi, Z., Babaee Tirkolaee, E., Mirzazadeh, A., Weber, G.W. (eds.) Logistics and Supply Chain Management. LSCM 2020. Communications in Computer and Information Science, vol. 1458. Springer, Cham (2021). https://doi.org/10.1007/978-3-030-89743-7_5
4. Hofmann, E., Marco, R.: Industry 4.0 and the current status as well as future prospects on logistics. Comput. Ind. **89**(2017):23–34 (2017)
5. De Koster, R., Stam, D., Balk, B.-M.: Accidents happen: the influence of safety-specific transformational leadership, safety consciousness, and hazard reducing systems on warehouse accidents. J. Oper. Manag. **29**, 753–765 (2011)
6. Christou, M.-D.: Analysis and control of major accidents from the intermediate temporary storage of dangerous substances in marshalling yards and port areas. J. Loss Prev. Process Ind. **12**, 109–119 (1999)
7. Reaidy, P.-J., Gunasekaran, A., Spalanzani, A.: Bottom-up approach based on Internet of Things for order fulfillment in a collaborative warehousing environment. Int. J. Prod. Econ. **159**, 29–40 (2015)
8. Giannikas, V., Lu, W., McFarlane, D., Hyde, J.: Product intelligence in warehouse management: a case study. Lect. Notes Comput. Sci. (Springer) **80**(62), 224–235 (2013)
9. Eruguz, A.-S., Jemai, Z., Sahin, E., Dallery, Y.: Applying the guaranteed-service model to a decentralized supply chain: impact on the cycle-service-level. In: IESM Conference, October 2013, Rabat, Morocco (2013)
10. Guerriero, F., Musmanno, R., Pisacane, O., Rende, F.: A mathematical model for the multi-levels product allocation problem in a warehouse with compatibility constraints. Math. Model. **37**, 4385–4398 (2013)
11. Trab, S., Bajic, E., Zouinkhi, A., Abdelkrim, M.-N., Chekir, H., Hadj Ltaief, R.: Product allocation planning with safety compatibility constraints in IoT-based warehouse. Procedia Comput. Sci. **73**(2015), 290–297 (2015)
12. Chiang, D.-M.-H., Lin, C.-P., Chen, M.-C.: The adaptive approach for storage assignment by mining data of warehouse management system for distribution centres. Enterp. Inf. Syst. **5**(2), 219–234 (2011)
13. Claudia, B., Andrea, F., Luigi, L., Massimiliano, M.: Multiproduct slot allocation heuristic to minimize storage space. Int. J. Retail Distrib. Manag. **42**(3), 172–186 (2014)
14. Klein, T., Thomas, A.: Opportunities to reconsider decision making processes due to Auto-ID. Int. J. Prod. Econ. **121**, 99–111 (2009)
15. Liu, Q., Liu, M., Wang, Z., et al.: A novel intelligent manufacturing mode with human-cyber-physical collaboration and fusion in the non-ferrous metal industry. Int. J. Adv. Manuf. Technol. (2021). https://doi.org/10.1007/s00170-021-08250-5

16. Wenyan, S., Xinguo, M., Pengpeng, W.: Collaborative product innovation network: status review, framework, and technology solutions. Concurr. Eng.: Res. Appl. **21**(1), 55–64 (2012)
17. Pannequin, R., Thomas, A.: Stigmergy: a design pattern for product-driven systems. In: IFAC Symposium on Information Control Problems in Manufacturing, Moscow, Russia (2009)
18. Wong, C.-Y., McFarlane, D., Zaharudin, A.-A., Agarwal, V.: The intelligent product driven supply chain. In: International Conference on Systems, Man And Cybernetics, 6–9 October, Hammamet, Tunisia (2002)
19. Morel, G., Panetto, H., Zaremba, M., Mayer, F.: Manufacturing enterprise control and management system engineering: paradigms and open issues. Annu. Rev. Control. **27**(2), 199–209 (2003)
20. Bajic, E.: Ambient services modeling framework for intelligent products. In: International Workshop on Smart objects systems, 11–14 September, Tokyo, Japan (2005)
21. De Koster, R., Le-Duc, T., Roodbergen, K.-J.: Design and control of warehouse order picking: a literature review. Eur. J. Oper. Res. **182**(2), 481–501 (2007)
22. Zouinkhi, A., Bajic, E., Rondeau, E., Ben Gayed, M.: Ambient intelligence: awareness context application in industrial storage. Wirel. Sens. Netw. **3**(3), 135–146 (2011)
23. Yang, L., Ye, M.: The design and development of intelligent warehouse management system based on .NET and internet of things. In: International Conference on Logistics, Engineering, Management and Computer Science, 24–26 May, Shenyang, China (2014)
24. Trab, S., Bajic, E., Zouinkhi, A., Abdelkrim, M.-N.: A hybrid decision-making-aided process for classified products warehousing. In: 15th International Multi-Conference on Systems, Signals & Devices (SSD), 19–22 March, Yasmine Hammamet, Tunisia (2018)
25. Brélaz, D.: New methods to color the vertices of a graph. Commun. ACM **22**(4) (1979)
26. Bandyopadhyay, A., Dhar, A.-K., Basu, S.: Graph coloring: a novel heuristic based on trailing path-properties, perspective and applications in structured networks. Soft Comput. **24**(2020), 603–625 (2020)
27. Metoui, B., Melchior, P., Najar, S., Abdelkrim, M.-N., Oustaloup, A.: Fractional attractive force for robust path planning. Int. J. Model. Identif. Control **15**(2), 134–143 (2012)
28. Khatib, O.: Real-time obstacle avoidance for manipulators and mobile robots. In: International Conference on Robotics and Automation, 25–28 March, St Louis, USA (1985)
29. Zbib, N., Pach, C., Sallez, Y., Trentesaux, D.: Heterarchical production control in manufacturing systems using the potential fields concept. J. Intell. Manuf. **23**(5), 1649–1670 (2012)
30. Trab, S., Bajic, E., Zouinkhi, A., Thomas, A., Abdelkrim, M.-N., Chekir, H., Hadj Ltaief, R.: A communicating object's approach for smart logistics and safety issues in warehouses. Concurr. Eng.: Res. Appl. J. **25**(1), 53–67 (2017)

Application of Artificial Intelligence (AI) in Supply Chain: An Overview

Laxmi Pandit Vishwakarma and Rajesh Kr Singh

Abstract Artificial Intelligence (AI) is the technology which is gaining popularity in the recent years. The application of AI can be seen in all industries. This article is focusing particularly on the application of AI in the supply chain. Looking at the numerous benefits of AI, many industries, researchers and practitioners across the world had started exploring this area. The AI consists of many tools such as machine learning, deep learning, expert systems, big data, artificial neural networks, speech processing, natural language processing, robotics etc. This study elaborates the applications of AI tools among different stakeholders of the supply chain. These stakeholders are the suppliers, manufacturers, logistics, retailers and the customers. Several benefits of AI have also been explored in this study through literature review. This study adds knowledge base in the field of technology adoption such as AI in the supply chain for sustainable development in emerging countries.

Keywords Artificial intelligence · AI · Supply chain

1 Introduction

Dash et al. [1] highlighted that "A well-functioning supply chain is a key to success for every business entity". Kumar et al. [2] observed that the Industry 4.0 technologies are taking the world to the next level and AI being one of the Industry 4.0 technology, provides several advantage. AI helps in solving day to day problems easily, with higher accuracy and speed. Kar et al. [3] examined AI as a powerful technology in today's world. The author highlights that the benefits and the applications of the AI is exponentially growing in current times. Dubey et al. [4] highlighted that the application of AI has recently gained attention in many industries. It has been

L. P. Vishwakarma (✉) · R. K. Singh
Management Development Institute, Gurgaon, India
e-mail: fpm21laxmi_p@mdi.ac.in

R. K. Singh
e-mail: rajesh.singh@mdi.ac.in

S. El Himer et al. (eds.), *Artificial Intelligence of Things for Smart Green Energy Management*, Studies in Systems, Decision and Control 446,
https://doi.org/10.1007/978-3-031-04851-7_12

observed that, broadly the application of AI is used across different sectors ranging from manufacturing to e-commerce [5]. Helo and Hao [6] revealed that with the presence of AI, the supply chain management will be changing drastically. Dwivedi et al. [7] and Mahroof [8] highlights that many big players like Philips, Amazon, eBay, Walmart etc. had already implemented AI in their supply chain due to its numerous benefits.

In today's competitive and fast changing environment, the supply chain needs to be resilient, provide accurate, rapid good and services with high levels of security and safety [9, 10]. The need for the adoption of new technologies arises when the existing supply chain models couldn't identify and respond towards the fluctuating demands, constantly changing the working of business processes and unstructured decision problems [11]. These problems need more advanced IT systems which can quickly take decisions and respond to the highly variable situations [12]. Firms that have focused and invested in building their supply chain flexible and resilient are the ones, which are not much affected by the disruptions [9, 13–16]. A resilient supply chain does not let the disruptions harm their operations and cash flow cycles.

Chunsheng et al. [11] highlighted that firms which are not adapting the change easily are facing higher level of difficulty in building resilience, responsiveness and restoration. Looking at the current COVID-19 pandemic situation, there are around 94% of the fortune 1000 companies around the world which has faced the disruption. Further there are around 75% of the top 1000 firms which have experienced negative effect on their business from the pandemic [17]. Ivanov and Dolgui [16] identified that organisations which are using intelligent network, are working better when compared to the organisations which has not yet adopted the intelligent networks.

The traditional supply chains were not much equipped with the planning that could easily assess the environment and make decisions for sourcing, distribution and scheduling the activities involved in a supply chain. But the adoption of new technologies like AI has made this possible and provides the ability to visualise and examine the risks which could occur in future. Self-learning and self-organising features of AI based technologies contributes to decision making process [18, 20]. Based on that, the firms now prioritise their action plans and take strategic decisions in relation to the inventory, capacity, logistics and compliances to meet the customer demand [18, 19]. The intelligent systems having the analytical capabilities and adequate information processing ensures the smart execution and performance of the supply chain [18].

Remaining chapter is organised as follows: Section two defines AI, section three deals with the applications of AI across supply chain, section four describes the benefits of AI application in supply chain and finally section five provides conclusion and future scope for study.

2 Definitions of AI

AI was introduced in the 1950s and came into attention during early 2000s, but till today there are many debates regarding the benefits and the consequences of using AI. Still, over the last two decades, the use of AI is continuously growing [21]. Dubey et al. [4] highlighted that AI is becoming popular and is considered as a thriving technology in the rapid changing technological landscape. The main reason for this is the increasing amount of data and complexity in different areas of business environment [22]. Several businesses found potential in AI technology and had used AI according to the needs and objectives of the organisation. Huang and Rust [23] defined AI as a system of multiple machines which are connected with each other and has the ability to depict human intelligence in planning, analysing, forecasting or proposing actions to address any business problem. AI empowers the business systems to work without human interventions. It learns from data and builds insights from them [20].

The "Association for the Advancement of AI (AAAI) defines AI as "advancing the scientific understanding of the mechanisms underlying thought and intelligent behaviour and their embodiment in machines". Soleimani [24] highlights that the major aim for AI is to create and to understand the intelligent entities. Schniederjans et al. [25] recognised that AI has the capability of solving problems quickly with higher levels of speed, accuracy and efficiency. Modgil et al. [18] encores that organisations are using the Artificial Intelligence (AI) based solutions to detect the potential concerns in the supply chain and to plan and execute further actions based on the situations. AI has been recognized as one of the significant technologies which enables smooth communication among multiple devices and machines [7, 26]. Russell and Norvig [27] and Soleimani [24] scientifically defined AI as the branch of intelligent agents which are rationally assessing the environment and are taking intelligent actions based on their assessments.

Duan et al. [28] examined AI as the ability of the machines to learn from past experiences and take decision on their own in a similar manner to the human with intelligence. Russell and Norvig [27] analysed the objective of AI as "to create rational agents who can perceive and act such that some objective function is optimised". Examples for the same are pattern recognition, machine vision, developing learning systems, natural processing systems and problem-solving systems for decision support. These developments are using deep neural network, mathematical practices, convolutional neural networks, optimization techniques and multiple numerical methods [6].

In simple terms, AI is the ability to learn new concepts either on its own or from the past experiences and to mimic human behaviour or human patterns. AI provides holistic benefits in supply chain management. It is mainly because this technology integrates big data, machine learning and deep learning [29, 30]. AI further consists of speech processing, vision processing, robots, expert systems, natural language processing, artificial neural networks, cyber-physical systems etc. [28].

- **Big data** reduces the redundancy throughout the supply chain and improves the decision-making capabilities of the firm. AI has the capacity of handling big data (Having high variety, volume, veracity, velocity, variability and value) [31]. Big data- heuristics algorithms help to extract knowledge from large amount of data collected from the sensors, operations and business transactions [32].
- **Machine learning** (ML) based systems helps in understanding the large data sets and then identify patterns from them. ML is mainly based on the gained knowledge through past experiences. It is an important tool in analysing customers and using such information for the organisation. This information can be used for product or process improvement or for new product development. For example, from customer feedback, the systems improve their problem-solving skills and performances [33]. ML directly contributes in strengthening the procurement strategy as it automates the contracts with the suppliers [21, 34].
- **Artificial neural networks** (ANN) work similarly as the human nervous system. As in a human nervous system, the nerves are connected. In a similar manner, the computer memories are interconnected with each other The information which are output from one neuron acts as an input for another neuron. These neurons are linked with each other and the weights assigned to them are responsible for the weakening or strengthening of the information passed through this linked neuron. This resulting in learning, recognising patterns, extracting information and clustering objects [35].
- **Expert systems** are developed on the basis of computer programs which handles large amount of data, analyses them and provides numerous options [36] which are further used by human experts of domain knowledge [37] to take decisions during times [21]. These systems are usually used to enhance the productivity and managing the logistics or warehousing. These expert systems are known for cost effectiveness, selection or evaluation of suppliers or third-party logistics providers, planning or designing products or processes and working on real time basis.
- **Speech Processing** (SP) are used in capturing data from voices or speeches. These systems transmit the data from speeches into speech digital signals [38]. Fu and Sun [38] and Levy [39] highlights that in these systems the users can also respond back to the system immediately and easily communicate with them.
- **Natural Language Processing** (NLP) systems extracts information from patterns which are in the form of either text or speech. These systems analyse phrases or words and identifies the relation between them [37].
- **Robotics** is the interdisciplinary branch which includes information, mechanical, computer and electronics engineering. The robots are used for performing work on their own. The use of robots helps in reducing human error as well as performing tasks faster [37].

3 Application of AI Across Supply Chain

Supply chain is the complex and integrated concept comprising suppliers, manufacturers, logistics, retailers and consumers [6]. Panetta [40] argued that the application of AI in supply chain is yet not completely explored. Seyedghorban et al. [12] highlighted that in today's world, there is a need for transparent supply chain in relation to the business stakeholders, structures and tasks. The present supply chain aspires to improve the responsiveness, provide quick delivery services, optimises the internal and external processes and networks, be agile and build resilience [40]. All these goals of the supply chain can be achieved by adopting the Industry 4.0 technology and AI is one of them. By digitalising the business processes, organisations are easily able to fulfil the customer demand. Further leading to gain competitive advantage [41]. AI helps the supply chain professionals to update their skill set and to take important decisions during emergencies in any normal working atmosphere as well as in a cross-cultural atmosphere [42–44].

The supply chain flow diagram is shown in Fig. 1. The stakeholders involved in the supply chain are the suppliers, manufacturers, logistics, retailers and the consumers. The supply chain processes are the purchasing, manufacturing, transportation, warehousing and the distribution.

3.1 Suppliers

In supply chain, make or buy decision is the primary concern for many organisations. Making the intermediate products or buying them from the external sources and utilising their resources to focus on their own competency is one of the major confusions which most of the companies have. The expert systems help the organisations in taking these decisions by generating "what-if" scenarios. These systems also automate the ordering process by developing hybrid agents or knowledge-based systems using AI. It further helps the organisation in timing bidding for their raw materials and screening their qualified suppliers. Evaluating online catalogues from multiple suppliers is yet another application of AI for suppliers in the supply chain management.

Muniz et al. [45] examined that AI helps in order fulfilment, inventory planning, demand planning and designing network in relation to the supplier's system. AI helps in material handling at the suppliers and the receiver's end. AI helps in building the procurement strategy by automating the contracts with the suppliers [21, 34]. Agents based systems in AI helps the firms in global sourcing.

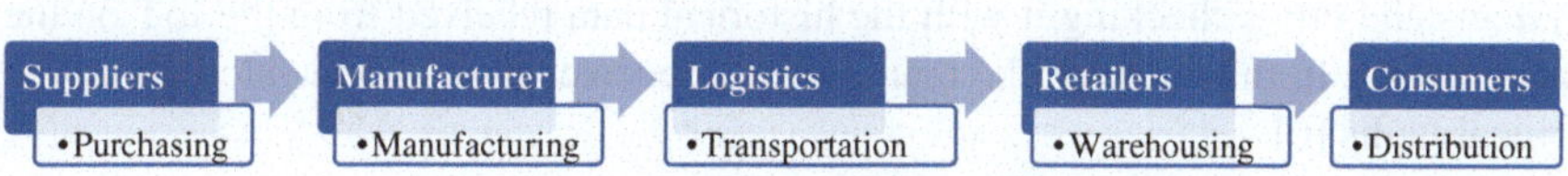

Fig. 1 Supply chain flow diagram. *Source* Helo and Hao [6] and Min [50]

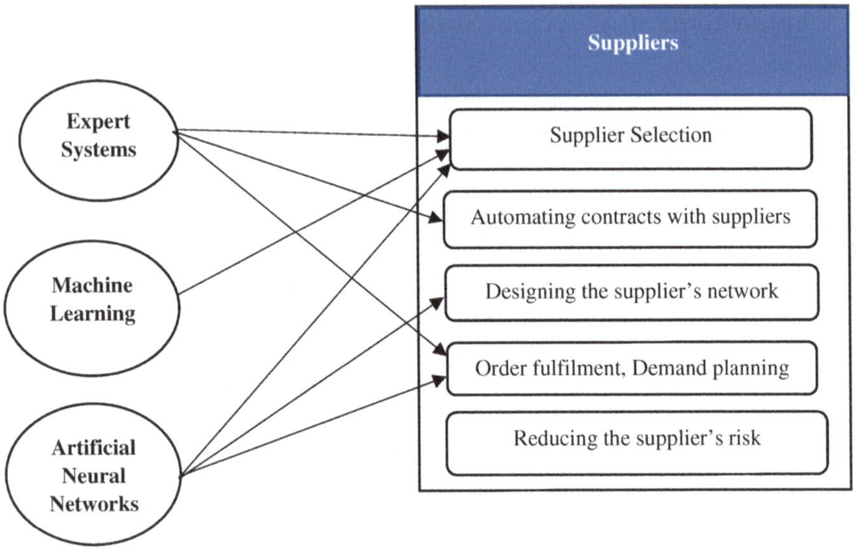

Fig. 2 Application of AI tools at the supplier stage of the supply chain. *Source* Helo and Hao [6] and Min [50]

Figure 2 states the application of tools (expert systems, machine learning and the artificial neural networks) of AI at the supplier stage of the supply chain.

3.2 Manufacturers

The advantage technologies have digitalised almost each stage of the manufacturing processes [46]. At multiple stages, managing inventory is a crucial task. Many organisations attain their competitive advantage and are able to save costs on the basis of their inventory management. The amount, size and quality of inventory required at the time of manufacturing, the order cycle, the accuracy, everything plays an equally important role in fulfilling customer orders while maintaining cost efficiency in the organisation. The AI helps in demand forecasting and thereby managing the inventory accordingly [18, 19]. AI has the capacity to capture the inventory patterns throughout different seasons and managing all levels of inventory in the supply chain without causing bullwhip effect [35, 47]. For example: ABB is collaborating with IBM Watson to obtain real time cognitive insights. The company is using AI to detect the defects in the assets by capturing the real time production images from the ABB systems and cross checking it with the historical data received from the IoT on the factory floor. In this manner, the manufacturer receives the real time alert about the critical faults [6].

Dash et al. [1] highlights that AI has the potential to provide 100 accuracies in forecasting customer demand. Having all the past data records on bill of material, production schedules, order patterns in the ERP system, the manufacturers gain knowledge about the consumption pattern, forecast demands and automates the service order with the help of expert systems, machine learning and artificial neural networks. There are many forecasting methods which are been used from quite a long time. These techniques are used for short term as well as for long term projections. The moving average method, exponential smoothing or the time series forecasting methods are some of the traditional methods used in forecasting. But these methods largely depend on the validity as well as the accuracy of the historic data. These methods provide better results for any existing products or services on which any historic data is applicable. But it fails when the forecasting is needed for any new product or services or for any innovative product or services.

In such cases, the AI is used for forecasting as AI considers and understand several other factors then the historical data and better predicts what products or services will be better for the consumers, what time will be better for the launch of the product or services, what age group to be target for the particular product or services, at what price the product or the services should be launched to the consumer etc. For example: Netflix uses AI in demand forecasting. It collects the data from its user and draws patterns based on their preferences and tastes. Netflix deeply observes their users' activities. For example, if a person named Evan watches A. and a different person named Ryan of similar taste and preferences watches B. Then, Netflix recommends B to Evan and A to Ryan. Based on this, Netflix is successfully engaging their customers and launching new shows and movies. AI further facilitates the preventive maintenance techniques and promotes time to time remote inspections of the machines. It uses the data from sensors, hardware, robots, drones or any machine learning application and smoothens the complex tasks and reduces the downtime of the machines [48, 49].

Figure 3 states the application of tools (expert systems, machine learning and the artificial neural networks) of AI at the manufacture stage of the supply chain.

3.3 Logistics

Due to globalization, supply chain has expanded its geographical reach and somewhere managing the logistics has been critical for many organisations. But, with the implementation of AI, organisations are now easily able to examine the traffic, weather and accordingly manages their routes. AI helps the firms to navigate different markets separately at macro levels and to take strategic decisions based on that [21]. The agent-based systems in AI helps in managing intra-organisational logistics. For example: A Japanese company named NEC has used AI in their logistics operations to verify the product shipped details with the product shipment schedule lists. This technique uses the unique image recognition technology rather than the general barcode method or any other information [51]. AI helps to select and evaluate the

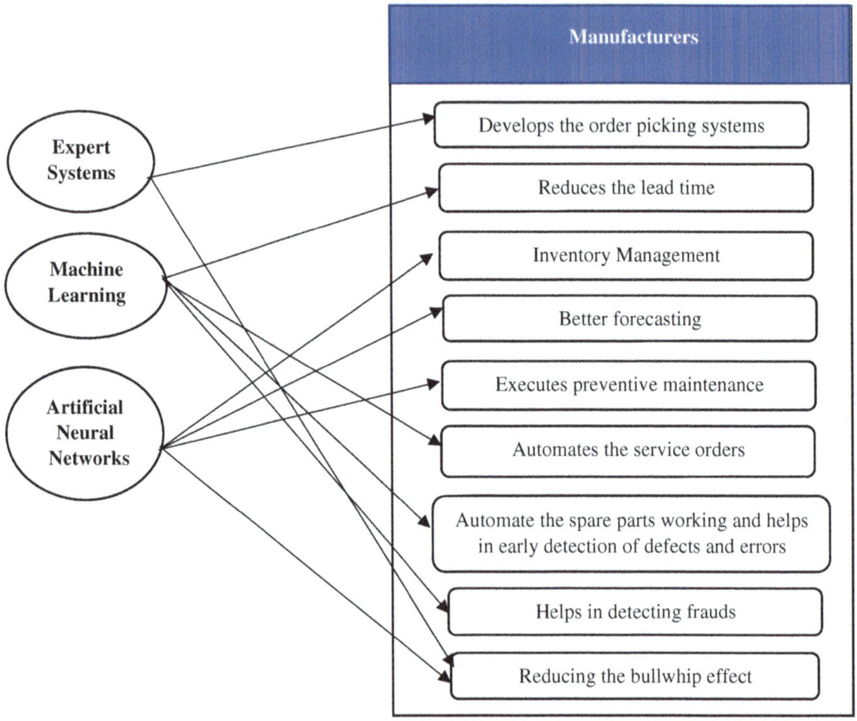

Fig. 3 Application of AI tools at the manufacturer stage of the supply chain. *Source* Helo and Hao [6] and Min [50]

third- party logistics provider. It helps the organisations to develop logistics strategy to fulfil customer's demand.

In a supply chain, while delivering the products, every minute and mile matters. Hence, efficiency and productivity of the firm plays an important role at this stage. Webster et al. [52] highlighted that AI-enabled logistics robots help to enhance the robustness of the processes. They integrate different routes and makes more precise movements for the robots. Vehicle routing, traffic assignment, space utilization are a few major problems which organisations are facing. AI helps to deal with such problems. AI driven robots and machines helps in boosting the productivity and efficiency of the supply chains. AI facilitates the shipment tracking and the risks associated with it [14]. The algorithms used in these machines and robots optimises routes and results in reducing time and cost incurred in the process [53].

Figure 4 states the application of tools (expert systems, machine learning, the artificial neural networks and robotics) of AI at the logistics stage of the supply chain.

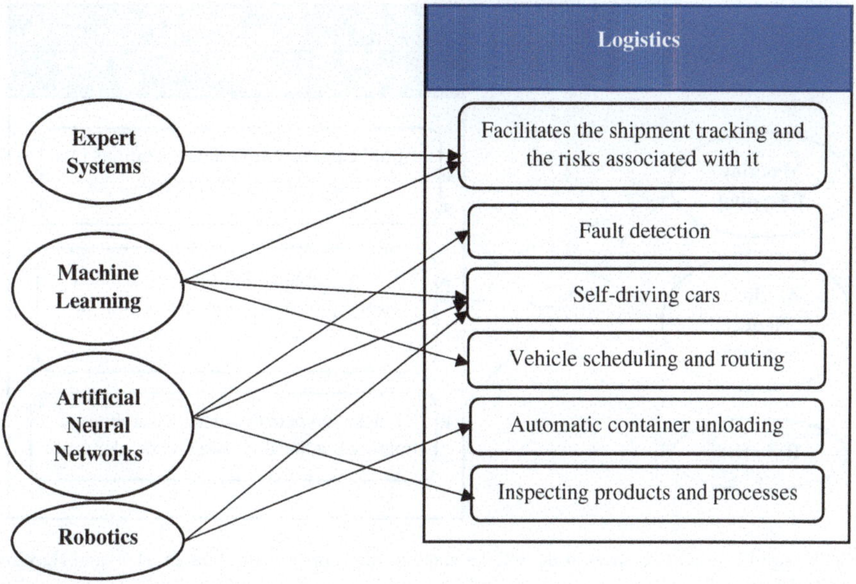

Fig. 4 Application of AI tools at the logistics stage of the supply chain. *Source* Helo and Hao [6] and Min [50]

3.4 Retailers

AI has now empowered the retailers to know more about what the customer wants even before the customer purchases something for themselves [54]. This is made possible through the forecasting done on the basis of the huge data volumes and patterns which the company draws from online viewing history, previous transactions, seasonal shopping patterns, social media trends, facial expression analysis, weather forecasts etc. [55, 56]. AI simplifies the warehouse operations by automating the process involved in it [57]. The barcode scanners at the retailers are been used for predictive analysis to forecast demand [14]. For example: A German company named Otto cut uses deep learning to analyse the past billions of transactions made by the customers and predicts what will be the next order or the preference of their customer is going to be. Based on the prediction, the companies surpass their stock by 20% and minimises their product return by more than two million items a year. With this method, the German company further locate items for easy access and optimises their storage space.

Figure 5 states the application of tools (expert systems, machine learning, the artificial neural networks and robotics) of AI at the retailer stage of the supply chain.

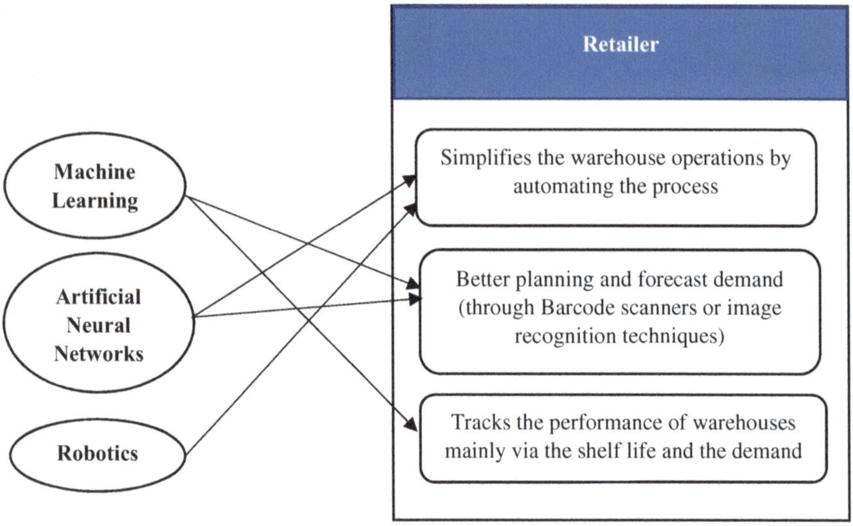

Fig. 5 Application of AI tools at the retailer stage of the supply chain. Source. Helo and Hao [6] and Min [50]

3.5 Consumers

To retain customers, the customer relationship management is constantly working with the organisation because in today's competitive world, the customers are having a variety of options in front of them and to retain them for the current product or service is a difficult task for the organisations. To improve and satisfy their customers, many organisations are adopting AI. Toorajipour et al. [58] and Wamba et al. [30] highlights that AI improves the quality of delivering the products and services to the customers. It further smoothens the customer complaints and queries via AI-driven chat boots. This results in manging the compliances easily at lower cost in a value supply chain network [59]. It assists in tracking the performance of warehouses mainly via the shelf life and the demand [60, 61]. AI examines the stock levels and assigns activities accordingly.

AI helps in engaging customers through personalized AI driven bots. Huang and Rust [57] identified that these bots assist the customer support team and helps in tracking the delivering of the products. Dash et al. [1] examines that AI helps the organisation in identifying target customers, understanding the demographic and the geographic preferences of the customers and based on that recommends the right product, at the right price and with the right message to their customers. For example: Walmart, one of the largest retailers is using AI to provide better services to their customer by analysing their activities on multiple social media platforms and gaining insights on their behaviour and preferences. Based on such information, Walmart is able to target their customers and directly send them the product's information [62].

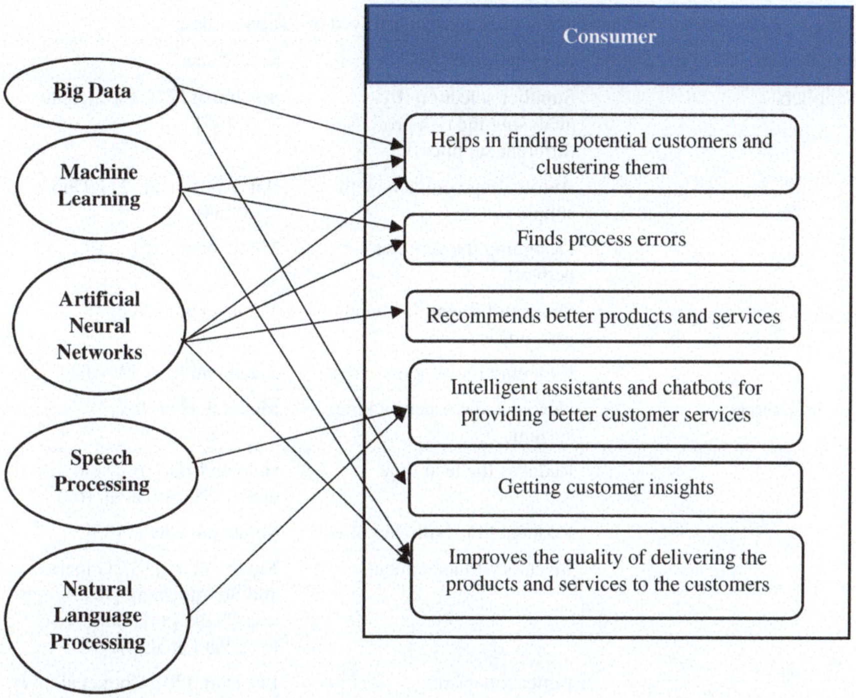

Fig. 6 Application of AI tools at the consumer stage of the supply chain. Source. Helo and Hao [6] and Min [50]

Sterne [63] illustrates the AI- supporting activities such as website operations, search engine optimization, making digital advertisements etc., which helps the customer to get better experience and let the organisations to interact closely with their customers. Deb et al. [54] and Gaggioli [64] highlighted that AI has revolutionized the search engines. Customers can now easily search the products either pressing a button or just by giving a voice command. The devices like Apple's Siri, Amazon's Echo etc. are used to perform such searches.

Figure 6 states the application of tools (big data, expert systems, machine learning, the artificial neural networks, speech processing and natural language processing) of AI at the consumer stage of the supply chain.

The Table 1 shows the summary of the application of AI among the stakeholders involved in a supply chain:

4 Benefits of AI Applications in a Supply Chain

The various benefits of AI in a supply chain are explained below:

Table 1 Application of AI among the stakeholders involved in a supply chain

Stakeholders in supply chain	Application of AI	References
Suppliers	Supplier selection (by analysing the risk profiles of different suppliers)	Soleimani [24], Cavalcante et al. [65]
	Automating contracts with suppliers	Baryannis et al. [21], Dubey et al. [34]
	Designing the supplier's network	Muniz et al. [45]
	Order fulfilment, Demand planning	Muniz et al. [45]
	Reducing the supplier's risk	Soleimani [24], Zhu et al. [66]
Manufacturers	Develops the order picking systems	Manzini, et al. [67]
	Reduces the lead time	Helo and Hao [6], Modgil et al. [18], Min et al. [68]
	Reducing the bullwhip effect	Singh and Challa [35]
	Inventory Management	Modgil et al. [18], Gunessee and Subramanian [19], Singh and Challa [35], Choi et al. [47], Paul et al. [69]
	Better forecasting	Lee et al. [20], Choi et al. [47], Amirkolaii et al. [70]
	Executes preventive maintenance	Lee et al. [20], Bughin et al. [48], Martin et al. [49]
	Improves the production capability of the firm	Helo and Hao [6]
	Automates the service orders from time to time	Helo and Hao [6], Hellingrath and Lechtenberg [71], Mortazavi et al. [72]
	Automate the spare parts working and helps in early detection of defects and errors in them	Helo and Hao [6], Lee et al. [20], Soleimani [24], Hellingrath and Lechtenberg [71], Mortazavi et al. [72]
	Helps in detecting frauds	Baryannis et al. [21], Davenport [37]
	Automate quality controls	Helo and Hao [6], Tellaeche and Arana [73], Taylan and Darrab [74]
	Automating call centres (services)	Brynjolfsson et al. [75]
Logistics	Facilitates the shipment tracking and the risks associated with it	Modgil et al. [14], Knoll et al. [79]

(continued)

Table 1 (continued)

Stakeholders in supply chain	Application of AI	References
	Self-driving cars	Brynjolfsson et al. [75]
	Vehicle scheduling and routing	Baryannis et al. [21], Levy [39], Choi et al. [47], Webster et al. [52], Wen et al. [53]
	Automatic container unloading	Webster et al. [52], Hengstler et al. [76]
	Inspecting products and processes	Jarrahi [36], Klumpp [77]
	Fault detection	Hengstler et al. [76]
Retailers	Simplifies the warehouse operations by automating the process	Huang and Rust [23], Klumpp [77]
	Better planning and forecast demand (through Barcode scanners or image recognition techniques)	Modgil, Singh and Hannibal [14]
	Tracks the performance of warehouses mainly via the shelf life and the demand	Butt [60], Queiroz et al. [61]
Consumer	Helps in finding potential customers and clustering them	Helo and Hao [6], Hellingrath and Lechtenberg [71], Knoll et al. [78]
	Finds process errors	Helo and Hao [6], Hengstler et al. [76]
	Recommends better products and services	Helo and Hao [6], Brynjolfsson and McAfee [33]
	Intelligent assistants and chatbots for providing better customer services	Dwivedi et al. [7], Modgil et al. [18], Gunessee and Subramanian [19]
	Getting customer insights	Choi et al. [47], Sterne [63]
	Improves the quality of delivering the products and services to the customers	Wamba et al. [30], Toorajipour et al. [58], Sterne [63]

i. **Strengthening supply chain resilience**: Wamba and Akter [80] observed that AI helps in identifying the weak supply chain nodes and strengthens supply chain resilience by working on such nodes. de Sousa Jabbour et al. [83] argued that to maintain a balance between the demand and supply during any stressful situations, the supply chains must configure their networks. AI helps in configuring the supply chain networks. The intelligent twins made via the AI- based mechanisms makes the above configuration easy and faster. The configurations made via the use of AI involves the supplier's supplier, suppliers, retailers and customers on the basis of their processing capabilities, the total costs involved in it and considering the lead time [18]. AI captures the day-to-day operational data and integrates different risks nodes, which further strengths the supply chain and make them more resilient [15]. This is mainly possible as by integrating the data, AI applies the what-if-scenarios and laid down guidelines which turns out to be as the contingency plans during the inevitable situations [31]. Ivanov [84] highlighted that organisation have now recognized the importance of the contingency plans, are constantly working to mitigate the supply shocks and are leading towards building a safer and effective supply chain. Modgil, Singh and Hannibal [14] observed the critical areas where AI is enhancing the supply chain resilience. These are mainly ensuring the last mile delivery, minimizing the negative affect of the disruptions, improving transparency, facilitating the agile procurement strategy and offering personalised solutions for the supply chain stakeholders.

ii. **Automate Business Processes**: AI makes the workforce creative by engaging them in the automate business processes [5]. AI automates and improves business process through continuous learning [1]. IT further reduces the manpower used to perform operations thereby reduces the human errors involved in them.

iii. **Optimising internal and external supply chain**: Dash et al. [1] revealed that a balance between the supply and demand is a critical factor for many organisations and optimisation helps to attain that. Jabbour et al. [5]; Soleimani [24] and Ivanov and Dolgui [16] revealed that AI helps in optimising the entire supply chain both internally and externally. It helps in planning and mapping the supply chain network. Dash et al. [1] added that AI along with optimising the supply chain, optimises the Research and Development also for many organisations.

iv. **Helps in sensing the market**: Dash et al. [1] highlighted that AI helps the organisation in sensing the market. It helps them to understand the customer preferences, identify the target customers, define the price, understand the demographic preferences and design the right message for them.

v. **Fulfilling customer expectation**: AI senses the market and provides the customer expectations on a faster rate [5]. It is mainly because it understands the consumption patterns or the consumers tastes and preferences based on the activities which the consumer performs in the social networking sites or on the bases of their history of the purchasing behaviour [30, 58].

vi. **Identifying business identity in supply chain**: AI helps the business in finding out their identify in the market, differentiating them from their competitors and working on their strengths [6]. At the same time, it helps the organisation to work on their weaknesses and innovations [80].

vii. **Investing in right practices**: AI provides suggestions to make the right investment practices [80]. AI helps to understand prior, which products or services could retain in the market based on applied analytics and thereby making investments in the right practices.

viii. **Help in decision- making process**: Decision making is always considered as an important step for any organisation. Usually, the decisions are taken by the senior level people in an organisation. But without data, the management will not be able to come to any conclusion [34]. The AI helps the organisation to extract information from the data and draws conclusion based on that, which furthers helps the management to arrive to any decision [7]. AI based algorithms help in decision- making process, ensures smart execution and performance of supply chain at all levels [85].

ix. **Provides real time coordination and collaboration**: AI provides real time coordination and collaboration to improve the visibility of the supply chain. Wamba et al. [30] highlighted that AI based innovations helps in real time collaboration and coordination, which further provides visibility among the entire supply chain network. Such collaborations and visibility improve the communication between the contractors and the suppliers during the stressful times [14]. Based on the visibility capability from AI based technology, supply chain professionals can easily generate insights on how the entire supply chain can be improved or changes can be made over a period of time [81]. Helo and Hao [6] highlights that AI increases the visibility and transparency across the supply chain.

x. **Contributes to sustainable development**: Gretzel et al. [82] highlights that AI helps in creating sustainable and optimal value in the supply chain. AI has encouraged the organisation to develop sustainable goods and services [86–88]. Duan et al. [28] examined that AI is helping the organisations to achieve the Sustainable Development Goals through technological innovations, provides positive effects on the environment and the civil society.

xi. **Helps in gaining competitive advantage**: AI helps to finds the strong points regarding their products or services for any organisation [7]. Once the organisation is able to understand their strength, they can market themselves according to them and easily differentiate themselves from their competitors. Thereby helps attaining competitive advantage [41].

xii. **Reduces Costs**: Dash et al. [1] examines that AI reduces the hidden costs in the manufacturing processes. Chui et al. [89] observes that AI has the potential of reducing the inventory cost by 5%. Panetta [40] highlights that with the implementation of intelligent technology in supply chain, organisations are able to get digital solutions, implement them in their supply chain and save money. The Table 2 summarises the benefits of AI in the supply chains.

5 Conclusion, Limitations and Future Scope

The supply chain is one of the most crucial parts of any organisation. Different stakeholders involved in the supply chain are the suppliers, manufacturers, logistics, retailers and customers. Fatorachian and Kazemi [94] identified critical success factors for the supply chain such as process integration, information sharing and collaboration. For all these factors, the supply chain should be made in such a way that it should collect the data in real time basis and should be digitally operated. All these parameters, are fulfilled by the adoption of AI in supply chain. Chui and

Table 2 Benefits of AI in the supply chain

Benefits of AI	References
Helps in integrating risk nodes and strengthening supply chain resilience	Jabbour et al. [5], Dwivedi et al. [7], Modgil et al. [14], Grida et al. [15], Ivanov and Dolgui [16], Modgil et al. [18], Wamba et al. [31], Wamba and Akter [80]
Automate business processes	Jabbour et al. [5], Helo and Hao [6], Baryannis et al. [21], Dubey et al. [34], Panetta [40], Huang and Rust [57]
Optimising internal and external supply chain	Dash et al. [1], Jabbour et al. [5], Helo and Hao [6], Ivanov and Dolgui [16], Modgil et al. [18], Soleimani [24]
Helps in sensing the market	Dash et al. [1], Jabbour et al. [5], Helo and Hao [6], Panetta [40]
Fulfilling customer expectation	Jabbour et al. [5], Helo and Hao [6], Wamba et al. [30], Toorajipour et al. [58]
Identifying business identity in supply chain	Helo and Hao [6], Wamba and Akter [80]
Investing in right practices	Wamba and Akter [80]
AI based algorithms help in decision-making process, ensures smart execution and performance of supply chain at all levels	Dwivedi et al. [7], Modgil et al. [14], Modgil et al. [18], Baryannis et al. [21], Wamba et al. [30], Dubey et al. [34], Panetta [40], Wamba and Akter [80], Tegmark [85], Zijm and Klumpp [95]
Provides real time coordination and collaboration to improve the visibility of the supply chain	Dash et al. [1], Helo and Hao [6], Modgil et al. [14], Wamba et al. [30], Panetta [40], Bock et al. [81]
Contributes to sustainable development	Gretzel et al. [82], Wirtz and Müller [86], Govindan et al. [87], Tuan et al. [88], Garbuio and Lin [91], Kuo and Smith [92], Merlino and Sproġe [93]
Helps in gaining competitive advantage	Dash et al. [1], Dwivedi et al. [7], Panetta [40], Tammela et al. [41]
Reduces costs	Dash et al. [1], Panetta [40]

Malhotra [90] found that there are many articles focusing on the application of AI in supply chain and has created the most value in the manufacturing industry. By integrating this technology, supply chain will be able to show the characteristics of self-governing, self- optimising, self- aware and self- determining.

In this study, the application of AI in the supply chain has been discussed. Considering all the stakeholders of the supply chain, the application of AI at each stage of supply chain has been elaborated. Small cases are discussed in the sub topics wherein, how the firms have implemented the AI and benefits attained. For example., in the manufacturing, ABB in collaboration with the IBM Watson has used the AI to detect the defects in the assets. The application of AI in demand forecasting by Netflix is further discussed. Under the Logistics section, a Japanese firm named NEC is using

AI to verify the product shipment details. Under the Retailer section, a German firm named Otto is applying AI to forecast the next order of their customers are written. AI is also helping the organisation like Walmart to gain consumer insights. Also, the study highlights the application of tools like machine learning, artificial neural networks, deep learning, robotics, speech processing, natural language processing, expert systems, big data in the supply chain.

Further, the study illustrated the numerous benefits of AI applications in supply chain. These benefits are AI are helping the organisations to integrate risk nodes and improving supply chain resilience, automating their business processes, optimising the internal as well as the external supply chain, helping in sensing the markets making investments in the right practices, helping in decision making process, providing them the real time coordination and collaboration to improve the visibility of the supply chain, thereby contributing to sustainable development. Findings of this study will help the organisations who have not yet adopted the AI in their supply chain. This chapter provides a brief summary of the benefits of AI and the application of AI in the supply chain, The adoption of AI in the supply chain is yet to be explored and adopted by many organisations. Findings of the study are based on review of relevant studies but findings cannot be generalised. More future empirical studies should be conducted in this field to examine these benefits and to extend the knowledge base. In the future, the identified benefits of the AI in the supply chain can be empirically tested and analysed.

References

1. Dash, R., McMurtrey, M., Rebman, C., Kar, U.K.: Application of artificial intelligence in automation of supply chain management. J. Strateg. Innov. Sustain. **14**(3), 43–53 (2019)
2. Kumar, V., Ramachandran, D., Kumar, B.: Influence of new-age technologies on marketing: a research agenda. J. Bus. Res. (2020). https://doi.org/ https://doi.org/10.1016/j.jbusres.2020.01.007
3. Kar, U.K., Dash, R.: The future of health and healthcare in a world of artificial intelligence. Arch. Biomed. Eng. Biotech. (2018)
4. Dubey, R., Gunasekaran, A., Childe, S.J., Bryde, D.J., Giannakis, M., Foropon, C., Hazen, B.T.: Big data analytics and artificial intelligence pathway to operational performance under the effects of entrepreneurial orientation and environmental dynamism: a study of manufacturing organisations. Int. J. Prod. Econ. **226**, 107599 (2019)
5. Jabbour, C.J.C., Fiorini, P.D.C., Ndubisi, N.O., Queiroz, M.M., Piato, E.L.: Digitally-enabled sustainable supply chains in the 21st century: a review and a research agenda. Sci. Total Environ. (2020). https://doi.org/10.1016/j.scitotenv.2020.138177
6. Helo, P., Hao, Y.: Artificial intelligence in operations management and supply chain management: an exploratory case study. Prod. Plan. Control 1–18 (2021)
7. Dwivedi, Y.K., Hughes, L., Ismagilova, E., Aarts, G., Coombs, C., Crick Galanos, T., et al.: Artificial Intelligence (AI): multidisciplinary perspectives on emerging challenges, opportunities, and agenda for research, practice and policy. Int. J. Inf. Manage. 101994 (2019)
8. Mahroof, K.: A human-centric perspective exploring the readiness towards smart warehousing: the case of a large retail distribution warehouse. Int. J. Inf. Manage. **45**, 176–190 (2019). https://doi.org/10.1016/j.ijinfomgt.2018.11.008

9. Zouari, D., Ruel, S., Viale, L.: Does digitalising the supply chain contribute to its resilience? Int. J. Phys. Distrib. Logist. Manag. **51**(2), 149–180 (2020). https://doi.org/10.1108/IJPDLM-01-2020-0038
10. Sharma, M., Luthra, S., Joshi, S., Kumar, A.: Developing a framework for enhancing survivability of sustainable supply chains during and post-COVID-19 pandemic. Int. J. Log. Res. Appl. (2020). https://doi.org/10.1080/13675567.2020.1810213
11. Chunsheng, L., Wong, C.W., Yang, C.C., Shang, K.C., Lirn, T.C.: Value of supply chain resilience: roles of culture, flexibility, and integration. Int. J. Phys. Distrib. Logist. Manag. **50**(1), 80–100 (2019)
12. Seyedghorban, Z., Tahernejad, H., Meriton, R., Graham, G.: Supply chain digitalization: past, present and future. Prod. Plan. Control **31**(2–3), 96–114 (2020). https://doi.org/10.1080/09537287.2019.1631461
13. Belhadi, A., Kamble, S., Jabbour, C.J.C., Gunasekaran, A., Ndubisi, N.O., Venkatesh, M.: Manufacturing and service supply chain resilience to the COVID-19 outbreak: lessons learned from the automobile and airline industries. Technol. Forecast. Soc. Chang. (2021). https://doi.org/10.1016/j.techfore.2020.120447
14. Modgil, S., Singh, R.K., Hannibal, C.: Artificial intelligence for supply chain resilience: learning from Covid-19. Int. J. Logistics Manage (2021)
15. Grida, M., Mohamed, R., Zaied, A.N.H.: Evaluate the impact of COVID-19 prevention policies on supply chain aspects under uncertainty. Transp. Res. Interdisc. Perspect. (2020). https://doi.org/10.1016/j.trip.2020.100240
16. Ivanov, D., Dolgui, A.: Viability of intertwined supply networks: extending the supply chain resilience angles towards survivability. A position paper motivated by COVID-19 outbreak. Int. J. Prod. Res. **58**(10), 2904–2915 (2020)
17. Accenture.com (2020) Rapid response: a pragmatic approach to maintaining supply chain resilience in times of uncertainty. Accenture-COVID19-Maintaining-Supply-ChainResilience-in-Times-of-Uncertainty.pdf (accessed 20 April 2021)
18. Modgil, S., Gupta, S., Stekelorum, R., Laguir, I.: AI technologies and their impact on supply chain resilience during COVID-19. Int. J. Phys. Distrib. Logist. Manage. (2021)
19. Gunessee, S., Subramanian, N.: Ambiguity and its coping mechanisms in supply chains lessons from the Covid-19 pandemic and natural disasters. Int. J. Oper. Prod. Manag. **40**(7–8), 1201–1223 (2020)
20. Lee, W.J., Wu, H., Yun, H., Kim, H., Jun, M.B., Sutherland, J.W.: Predictive maintenance of machine tool systems using artificial intelligence techniques applied to machine condition data. Procedia CIRP **80**, 506–511 (2019). https://doi.org/10.1016/j.procir.2018.12.019
21. Baryannis, G., Validi, S., Dani, S., Antoniou, G.: Supply chain risk management and artificial intelligence: state of the art and future research directions. Int. J. Prod. Res. **57**(7), 2179–2202 (2019)
22. Scholten, K., Scott, P.S., Fynes, B.: Mitigation processes–antecedents for building supply chain resilience. Supply Chain Manage. Int. J. **19**(2), 211–228 (2014)
23. Huang, M.H., Rust, R.T.: Artificial intelligence in service. J. Serv. Res. **21**(2), 155–172 (2018)
24. Soleimani, S.: A perfect triangle with: artificial intelligence, supply chain management, and financial technology. Arch. Bus. Res. **6**(11), 5681 (2018). https://doi.org/10.14738/abr.611.5681
25. Schniederjans, D.G., Curado, C., Khalajhedayati, M.: Supply chain digitisation trends: an integration of knowledge management. Int. J. Prod. Econ. **220**, 107439 (2020)
26. Guzman, A.L., Lewis, S.C.: Artificial intelligence and communication: a human-machine communication research agenda. New Media Soc. **22**(1), 70–86 (2020)
27. Russell, S.J., Norvig, P.: Artificial intelligence: a modern approach. Pearson Educ Limited, Malaysia (2016)
28. Duan, Y., Edwards, J.S., Dwivedi, Y.K.: Artificial intelligence for decision making in the era of Big Data–evolution, challenges and research agenda. Int. J. Inf. Manage. **48**, 63–71 (2019). https://doi.org/10.1016/j.ijinfomgt.2019.01.021

29. Gupta, S., Modgil, S., Bhattacharyya, S., Bose, I.: Artificial intelligence for decision support systems in the field of operations research: review and future scope of research. Ann. Oper. Res. (2021). https://doi.org/10.1007/s10479-020-03856-6
30. Wamba, S.F., Queiroz, M.M., Wu, L., Sivarajah, U.: Big data analytics-enabled sensing capability and organizational outcomes: assessing the mediating effects of business analytics culture. Ann. Oper. Res. (2020). https://doi.org/10.1007/s10479-020-03812-4
31. Wamba, S.F., Dubey, R., Gunasekaran, A., Akter, S.: The performance effects of big data analytics and supply chain ambidexterity: the moderating effect of environmental dynamism. Int. J. Prod. Econ. (2020). https://doi.org/10.1016/j.ijpe.2019.09.019
32. Kolinski, A., Horzela, A., Cudzilo, M., Domanski, R.: Reference model of information flow in business relations with 4pl operator. In: Kolinski, A., Dujak, D., Golinska-Dawson, P. (eds.), Integration of Information Flow for Greening Supply Chain Management, pp. 19–45. Springer, Cham (2020)
33. Brynjolfsson, E., McAfee, A.: Artificial intelligence, for real. Harv. Bus. Rev. (2017). https://hbr.org/2018/01/artificial-intelligencefor-the-real-world
34. Dubey, R., Gunasekaran, A., Childe, S.J., Bryde, D.J., Giannakis, M., Foropon, C., Hazen, B.T.: Big data analytics and artificial intelligence pathway to operational performance under the effects of entrepreneurial orientation and environmental dynamism: a study of manufacturing organizations. Int. J. Prod. Econ. (2020). https://doi.org/10.1016/j.ijpe.2019.107599
35. Singh, L.P., Challa, R.T.: integrated forecasting using the discrete wavelet theory and artificial intelligence techniques to reduce the bullwhip effect in a supply chain. Glob. J. Flex. Syst. Manag. **17**(2), 157–169 (2016). https://doi.org/10.1007/s40171-015-0115-z
36. Jarrahi, M.H.: Artificial intelligence and the future of work: human-ai symbiosis in organizational decision making. Bus. Horiz. **61**(4), 577–586 (2018). https://doi.org/10.1016/j.bushor.2018.03.007
37. Davenport, T.H.: From analytics to artificial intelligence. J. Bus. Anal. **1**(2), 73–78 (2018). https://doi.org/10.1080/2573234X.2018.1543535
38. Fu, T., Sun, B.: Application of speech recognition technology in logistics selection system. In: International Conference on Human Centred Computing, pp. 654–659. Springer, Cham (2017)
39. Levy, F.: Computers and populism: artificial intelligence, jobs, and politics in the near term. Oxf. Rev. Econ. Policy **34**(3), 393–417 (2018). https://doi.org/10.1093/oxrep/gry004
40. Panetta, K.: Gartner Predicts 2019 for Supply Chain Operations. Smarter with Gartner (2018). https://www.gartner.com/smarterwithgartner/gartner-predicts-2019-for-supply-chain-operations/
41. Tammela, I., Canen, A.G., Helo, P.: Time-based competition and multiculturalism: a comparative approach to the Brazilian, Danish and Finnish furniture industries. Manag. Decis. **46**(3), 349–364 (2008). https://doi.org/10.1108/00251740810863834
42. Dubey, R., Gunasekaran, A., Childe, S.J., Papadopoulos, T.: Skills needed in supply chain human agency and social capital analysis in third party logistics. Manag. Decis. **56**(1), 143–159 (2018)
43. Dubey, R., Gunasekaran, A.: Education and training for successful career in big data and business analytics. Ind. Commer. Train. **47**(4), 174–181 (2015)
44. Dubey, R., Gunasekaran, A., Altay, N., Childe, S.J., Papadopoulos, T.: Understanding employee turnover in humanitarian organizations. Ind. Commer. Train. **48**(4), 208–214 (2016)
45. Muniz, L.R., Conceiç~ao, S.V., Rodrigues, L.F., de Freitas Almeida, J.F., Affonso, T.B.: Spare parts inventory management: a new hybrid approach. Int. J. Logist. Manage. **32**(1), 40–67 (2020). https://doi.org/10.1108/IJLM-12-2019-0361.
46. Schiavone, F., Sprenger, S.: Operations management and digital technologies. Prod. Plan. Control **28**(16), 1281–1283 (2017). https://doi.org/10.1080/09537287.2017.1375151
47. Choi, T.M., Wallace, S.W., Wang, Y.: Big data analytics in operations management. Prod. Oper. Manag. **27**(10), 1868–1883 (2018). https://doi.org/10.1111/poms.12838
48. Bughin, J., Hazan, E., Ramaswamy, S., Chui, M., Allas, T., Dahlstrom, P., ... Trench, M.: Artificial intelligence: The next digital frontier? (2017)

49. Martin, C., Leurent, H.: Technology and innovation for the future of production: accelerating value creation. In: World Economic Forum, Geneva Switzerland (2017)
50. Min, H.: Artificial intelligence in supply chain management: theory and applications. Int. J. Log. Res. Appl. **13**(1), 13–39 (2010)
51. Umeda, Y., Muto, H., Tomita, M., Kondoh, K., Kominami, T., Hidaka, Y.: Warehouse product inspection system achieves work efficiency and quality improvements. NEC Tech. J. **12**(1), 40–44 (2017)
52. Webster, C., Ivanov, S.: Robotics, artificial intelligence, and the evolving nature of work. In: Digital Transformation in Business and Society, pp. 127–143. Palgrave Macmillan, Cham (2020)
53. Wen, J., He, L., Zhu, F.: Swarm robotics control and communications: imminent challenges for next generation smart logistics. IEEE Commun. Mag. **56**(7), 102–107 (2018)
54. Deb, S.K., Jain, R., Deb, V.: Artificial intelligence—creating automated insights for customer relationship management. In 2018 8th International Conference on Cloud Computing, Data Science & Engineering (Confluence), pp. 758–764. IEEE
55. Fildes, R., Ma, S., Kolassa, S.: Retail forecasting: research and practice. Int. J. Forecast. (2019)
56. Burgess, A.: AI in action. In: The Executive Guide to Artificial Intelligence. Palgrave Macmillan, Cham (2018)
57. Huang, M.H., Rust, R.T.: Engaged to a robot? the role of AI in service. J. Serv. Res. **24**(1), 30–41 (2020). https://doi.org/10.1177/1094670520902266
58. Toorajipour, R., Sohrabpour, V., Nazarpour, A., Oghazi, P., Fischl, M.: Artificial intelligence in supply chain management: a systematic literature review. J. Bus. Res. **122**, 502–517 (2021)
59. Treleaven, P., Batrinca, B.: Algorithmic regulation: automating financial compliance monitoring and regulation using AI and blockchain. J. Financ. Trans. **45**, 14–21 (2017)
60. Butt, A.S.: Strategies to mitigate the impact of COVID-19 on supply chain disruptions: a multiple case analysis of buyers and distributors. Int. J. Logist. Manag. (2021). https://doi.org/10.1108/IJLM-11-2020-0455
61. Queiroz, M.M., Ivanov, D., Dolgui, A., Fosso Wamba, S.: Impacts of epidemic outbreaks on supply chains: mapping a research agenda amid the COVID-19 pandemic through a structured literature review. Ann. Oper. Res. (2020). https://doi.org/10.1007/s10479-020-03685-7
62. Roden, S., Nucciarelli, A., Li, F., Graham, G.: Big data and the transformation of operations models: a framework and a new research agenda. Prod. Plan. Control **28**(11–12), 929–944 (2017). https://doi.org/10.1080/09537287.2017.1336792
63. Sterne, J.: From programming to statistics to machine learning for marketing. Appl. Market. Anal. **3**(4), 298–305(8) (2018)
64. Gaggioli, A.: Virtual personal assistants: an emerging trend in artificial intelligence. Cyberpsychol. Behav. Soc. Netw. **21**(12), 803–804 (2018)
65. Cavalcante, I.M., Frazzon, E.M., Forcellini, F.A., Ivanov, D.: A supervised machine learning approach to data-driven simulation of resilient supplier selection in digital manufacturing. Int. J. Inf. Manage. **49**, 86–97 (2019). https://doi.org/10.1016/j.ijinfomgt.2019.03.004
66. Zhu, Y., Zhou, L., Xie, C., Wang, G.J., Nguyen, T.V.: Forecasting SMEs' credit risk in supply chain finance with an enhanced hybrid ensemble machine learning approach. Int. J. Prod. Econ. **211**, 22–33 (2019). https://doi.org/10.1016/j.ijpe.2019.01.032
67. Manzini, R., Gamberi, M., Regattieri, A.: Design and Control of a Flexible Order-Picking System (FOPS) a new integrated approach to the implementation of an expert system. J. Manuf. Technol. Manag. **16**(1), 18–35 (2005). https://doi.org/10.1108/17410380510574068
68. Min, Q., Lu, Y., Liu, Z., Su, C., Wang, B.: Machine learning based digital twin framework for production optimization in petrochemical industry. Int. J. Inf. Manage. **49**, 502–519 (2019). https://doi.org/10.1016/j.ijinfomgt.2019.05.020
69. Paul, S.K., Azeem, A., Ghosh, A.K.: Application of adaptive neuro-fuzzy inference system and artificial neural network in inventory level forecasting. Int. J. Bus. Inf. Syst. **18**(3), 268–284 (2015). https://doi.org/10.1504/IJBIS.2015.068164
70. Amirkolaii, K.N., Baboli, A., Shahzad, M.K., Tonadre, R.: Demand forecasting for irregular demands in business aircraft spare parts supply chains by using artificial intelligence (AI). IFACPapersOnLine **50**(1), 15221–15226 (2017). https://doi.org/10.1016/j.ifacol.2017.08.2371

71. Hellingrath, B., Lechtenberg, S.: Applications of artificial intelligence in supply chain management and logistics: focusing onto recognition for supply chain execution. In: Bergener, K., Rackers, M., Stein, A. (eds.) The Art of Structuring, pp. 283–296. Springer, Cham (2019)
72. Mortazavi, A., Khamseh, A.A., Azimi, P.: Designing of an intelligent self-adaptive model for supply chain ordering management system. Eng. Appl. Artif. Intell. **37**, 207–220 (2015). https://doi.org/10.1016/j.engappai.2014.09.004
73. Tellaeche, A., Arana, R.: Machine learning algorithms for quality control in plastic molding industry. In: 2013 IEEE 18th Conference on Emerging Technologies & Factory Automation (ETFA), pp. 1–4 (2013)
74. Taylan, O., Darrab, I.A.: Fuzzy control charts for process quality improvement and product assessment in tip shear carpet industry. J. Manuf. Technol. Manag. **23**, 402–420 (2012). https://doi.org/10.1108/17410381211217434
75. Brynjolfsson, E., Rock, D., Syverson, C.: Artificial intelligence and the modern productivity paradox: a clash of expectations and statistics. In: The Economics of Artificial Intelligence: An Agenda. Chicago: University of Chicago Press (2018)
76. Hengstler, M., Enkel, E., Duelli, S.: applied artificial intelligence and trust—the case of autonomous vehicles and medical assistance devices. Technol. Forecast. Soc. Chang. **105**, 105–120 (2016). https://doi.org/10.1016/j.techfore.2015.12.014
77. Klumpp, M.: Automation and artificial intelligence in business logistics systems: human reactions and collaboration requirements. Int. J. Log. Res. Appl. **21**, 224–242 (2018). https://doi.org/10.1080/13675567.2017.1384451
78. Knoll, D., Neumeier, D., Pruglmeier, M., Reinhart, G.: An automated packaging planning approach using machine learning. Procedia CIRP **81**, 576–581 (2019). https://doi.org/10.1016/j.procir.2019.03.158
79. Knoll, D., Prüglmeier, M., Reinhart, G.: Predicting future inbound logistics processes using machine learning. Procedia CIRP **52**, 145–150 (2016). https://doi.org/10.1016/j.procir.2016.07.078
80. Wamba, S.F., Akter, S.: Understanding supply chain analytics capabilities and agility for data-rich environments. Int. J. Oper. Prod. Manag. **39**(6–8), 887–912 (2019)
81. Bock, D.E., Wolter, J.S., Ferrell, O.C.: Artificial intelligence: disrupting what we know about services. J. Serv. Mark. **34**(3), 317–334 (2020)
82. Gretzel, U., Sigala, M., Xiang, Z., Koo, C.: Smart tourism: Foundations and developments. Electron. Mark. **25**(3), 179–188 (2015). https://doi.org/10.1007/s12525-015-0196-8
83. de Sousa Jabbour, A.B.L., Jabbour, C.J.C., Hingley, M., Vilalta-Perdomo, E.L., Ramsden, G., Twigg, D.: Sustainability of supply chains in the wake of the coronavirus (COVID-19/SARS-CoV-2) pandemic: lessons and trends. Mod. Supply Chain Res. Appl. **2**(3), 117–122 (2020)
84. Ivanov, D.: Viable supply chain model: integrating agility, resilience and sustainability perspectives—lessons from and thinking beyond the COVID-19 pandemic. Ann. Oper. Res. (2020). https://doi.org/10.1007/s10479-020-03640-6
85. Tegmark, M.: Life 3.0: Being Human in the Age of Artificial Intelligence. Vintage Books, Division of Penguin Random House, New York, NY (2018)
86. Wirtz, B.W., Müller, W.M.: An integrated artificial intelligence framework for public management. Pub. Manage. Rev. **21**(7), 1076–1100 (2019). https://doi.org/10.1080/14719037.2018.1549268
87. Govindan, K., Jafarian, A., Nourbakhsh, V.: Designing a sustainable supply chain network integrated with vehicle routing: a comparison of hybrid swarm intelligence metaheuristics. Comput. Oper. Res. **110**, 220–235 (2019). https://doi.org/10.1016/j.cor.2018.11.013
88. Tuan, M.N.D., Thanh, N.N., Le Tuan, L.: Applying a mindfulness-based reliability strategy to the Internet of Things in healthcare–a business model in the Vietnamese market. Technol. Forecast. Soc. Change **140**, 54–68 (2019). https://doi.org/10.1016/j.techfore.2018.10.024
89. Chui, M., Henke, N., Miremadi, M. Most of AI's business uses will be in two areas. Harv. Bus. Rev. **20** (2018)

90. Chui, M., Malhotra, S.: AI Adoption Advances, but Foundational Barriers Remain. Mckinsey & Company, New York (2018)
91. Garbuio, M., Lin, N.: Artificial intelligence as a growth engine for health care startups: emerging business models. Calif. Manage. Rev. **61**(2), 59–83 (2019). https://doi.org/10.1177/0008125618811931
92. Kuo, T.C., Smith, S.: A systematic review of technologies involving eco-innovation for enterprises moving towards sustainability. J. Clean. Prod. **192**, 207–220 (2018). https://doi.org/10.1016/j.jclepro.2018.04.212
93. Merlino, M., Sproģe, I.: The augmented supply chain. Procedia Eng. **178**, 308–318 (2017). https://doi.org/10.1016/j.proeng.2017.01.053
94. Fatorachian, H., Kazemi, H.: Impact of industry 4.0 on supply chain performance. Prod. Plan. Control **32**(1), 1–19 (2020)
95. Zijm, H., Klumpp, M.: Logistics and supply chain management: developments and trends. In: Zijm, H., Klumpp, M., Clausen, U. (eds.) Logistics and Supply Chain Innovation, pp 1–20. Springer, Cham (2016)

Acorn Waste Valorization as Reinforcement in Polystyrene Composite: A Comparative Study

Maria Jabri, Oumayma Oulidi, Ilyasse Loulidi, Abdelouahed Amar, Abderahim Kali, and Fatima Boukhlifi

Abstract Today, plastics occupy a predominant place in the world's industrial production of which polystyrene (PS) is one of the most abundant plastics. In the context of sustainable industrial technology, scientific research favors the elaboration of plastics reinforced by organic waste, considering their low cost and their very interesting thermal and mechanical properties that will benefit the matrix. In the present study, we focused on acorn as bio-reinforcement, the PS was reinforced by the acorn pericarp (AP), and the acorn cupule (AC), separately. The results obtained show an integration of the reinforcement in the matrix, the FTIR spectra show the appearance of the OH band, the diffractograms show a decrease in intensity and thus a slight decrease in crystallinity, which remains very acceptable. Morphologically, the SEM images showed laminates for AP while the incorporation of AC caused pores in the matrix.

Keywords Polystyrene · Composite · Acorn waste · In situ polymerization

1 Introduction

Nowadays, plastic materials occupy a large place in the world's industrial productions, see their application field in a diverse area. Polystyrene (PS) is among the most abundant plastic materials due to its very interesting properties, especially in the field of textiles. As a result, this polymer has known a progressive production, which has led to an accumulation less tolerant to the environment. This has led to scientific research into the manufacture of composite materials reinforced with bio-reinforcement, on the one hand, the final material would be environmentally friendly

M. Jabri (✉) · I. Loulidi · A. Amar · A. Kali · F. Boukhlifi
Laboratory of Chemistry/Biology Applied to the Environment, Faculty of Sciences, Moulay Ismail University, url-cnrst-n°13BP 11201-Zitoune, 50070 Meknes, Morocco
e-mail: ma.jabri@edu.umi.ac.ma

O. Oulidi
Laboratory of Bio-Organic Chemistry, Molecular Materials and Environment, Faculty of Sciences, Moulay Ismail University, BP 11201-Zitoune, 50070 Meknes, Morocco

S. El Himer et al. (eds.), *Artificial Intelligence of Things for Smart Green Energy Management*, Studies in Systems, Decision and Control 446,
https://doi.org/10.1007/978-3-031-04851-7_13

Fig. 1 Diagram of the anatomy of an acorn (right): **a** Cupule, **b** Pericarp, **c** Integument, **d** Cotyledons, **e** Plantlet, **f** Radicle, **g** Remains of the style. **d**, **e**, and **f** form the embryo

and on the other hand less expensive since the reinforcement allows to minimize a significant amount of the matrix. In addition, the properties of the waste material are also of benefit, as it is a very considerable source of cellulose materials. Several lignocellulosic fibrous materials have been used to produce plastic composites as fillers or reinforcements.

Morocco has the subsoil of 294,378 ha or 15% of the world's subsoil. More than two-thirds of the latter are located in Portugal (29%), Spain (22%), and Algeria (21%). The rest (12%) is divided equally between Tunisia, France, and Italy. The subarray is highly valued for the diversity of its products (Wood, cork, sweet acorns, honey, mushrooms, medicinal plants, picnic area...), and offers important protection of the soil thanks to the permanent plant cover, which makes it a tree of great socio-economic and ecological importance [1].

Acorns are rich in amylopectin, which makes them less digestible (Fig. 1). In addition, acorn seeds are rich in low-toxicity tannins that are difficult to dispose of. As a result, many acorn resources are discarded each year. Acorn pericarp (AP) is rich in cellulose and lignin and compared to non-food plant fibers, AP has advantages such as high mechanical strength, resistance to decay, high chemical stability, and high hydrophobicity [2], find their application with success in several matrices such as Polyethylene [3] and reinforced epoxy resin. Within the framework of a sustainable industry, PS/AP and PS/AC composites were prepared by in situ polymerizations, with the mass percentages: 1, 5, and 10%. The impact of the incorporation of each of the two reinforcements AP and AC on the structural, thermal, and morphological properties was also studied.

2 Materials and Methods

2.1 Materials

Matrix: Styrene monomer, is utilized for preparing the polymer composite.

Reagents: Methanol.
Toluene.
Benzoyl peroxide (BPO).

2.2 Synthesis of Polystyrene

The polystyrene resin was synthesized by in situ polymerization of styrene in the presence of benzoyl peroxide and toluene. The reaction mixture was heated to 80 °C, and the temperature was kept constant for one hour. Afterward, the resin was allowed to cool to room temperature for 10 min, then washed with ethanol, filtered, and transferred to silicone molds.

2.3 Synthesis of Composites

the synthesis of the composites is done by the same protocol of pure polystyrene except that we add at the beginning of the synthesis the particles of reinforcements (1, 5, 10%) by weight.

2.4 Particles from Acorn Pericarp and Cupule

The samples of acorns were collected in the holm oak forest located in El-Hajeb, Morocco. The Acorn pericarp (AP) and cupule (AC) were washed, dried for 24 h at a temperature of 110 °C in an oven, then crushed and sieved to obtain fine, homogeneous samples (<75 μm). It's useful to know the composition of the reinforcement to better understand the behaviors observed during the inclusion of acorn pericarp and cupule into polystyrene. The investigations were conducted. The analyses were carried out using Shiguang's technique [4]. In (Table 1) the results of the constitutional analysis are shown.

Table 1 Constitutional analysis of Acorn pericarp and cupule

Particles reinforcement	Hemicellulose	Lignin	Cellulose	Extractives
Acorn pericarp	46.55	25.83	22.62	5
Acorn cupule	56.18	36.24	2.82	4.75

3 Characterization

3.1 *Fourier Transform Infrared (FTIP)*

The infrared spectra were obtained using a Fourier transform spectrometer (FTIR) of the JASCO 4000 type, equipped with a detector (TGS) and a ceramic source, separated by an optical system using an interferometer. Michelson. Samples (3% KBr) were packaged as pellets 12 mm in diameter. Absorption spectra were recorded in the range of 400 and 4000 cm^{-1} in absorbance or transmittance, with a resolution of 4 cm^{-1}.

3.2 *The XRD in Powder*

X-ray diffraction (XRD) patterns of PS composites were carried out at room temperature on a Philips instrument, using CuKα ($\lambda = 1.5406$ Å) radiation. The diffraction data were recorded in the 2θ range of 5–70° with a step of 0.04° at 40 kV and 20 mA.

3.3 *The Thermal Stability Was Determined Using TGA/the Thermogravimetric Analysis*

The thermal differential thermogravimetric ATG analyzes were carried out using a Shimadzu TA-60 type apparatus, operating under air with a linear heating rate of 10 °C/min from room temperature to room temperature 600 °C.

3.4 *MEB Analysis, the Surface Morphology of PS/AP, AC*

The surface morphology of PS composites was monitored by scanning electron microscope (SEM) using a Zeiss Sigma instrument at 3-keV acceleration voltage.

4 Results and Discussions

4.1 Effect on Morphology: Scanning Electron Microscopy (SEM)

SEM was used to study the morphological and structural properties of polystyrene and acorn pericarp and cupule (Fig. 2), Polystyrene/Acorn pericarp composites (Fig. 3), and Polystyrene/Acorn cupule composites (Fig. 4).

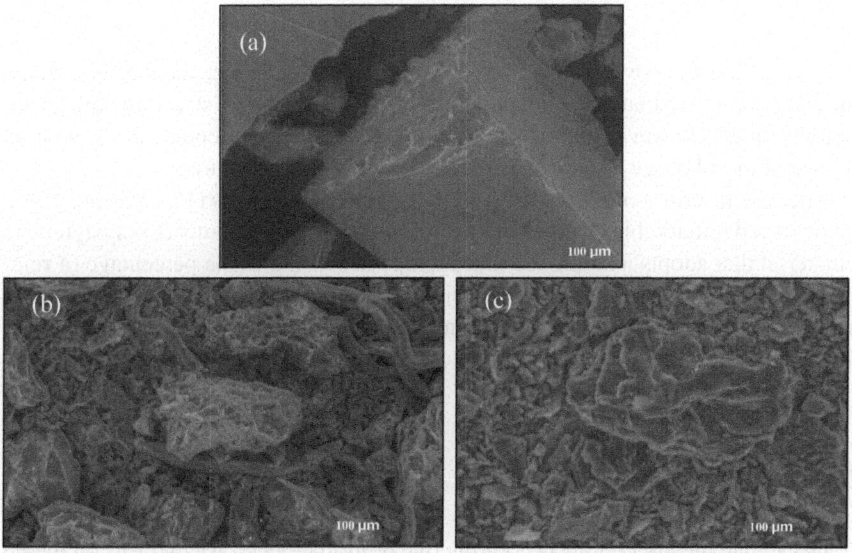

Fig. 2 SEM images of polystyrene and Particles from Acorn Pericarp and cupule, **a** PS, **b** AP, **c** AC

Fig. 3 SEM images of Polystyrene/Acorn pericarp composites, **d** PS + 1% AP; **e** PS + 5% AP; **f** PS + 10%AP

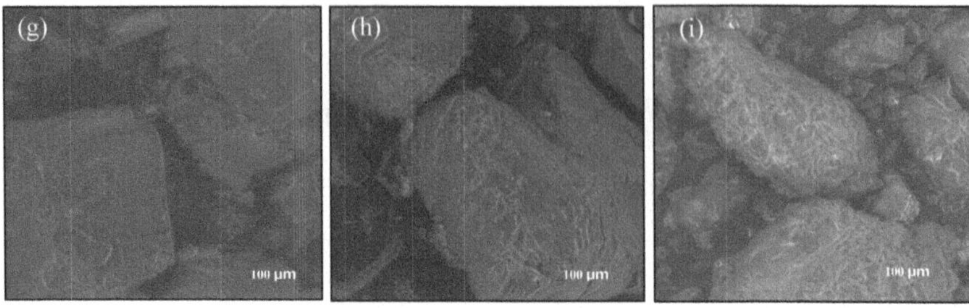

Fig. 4 SEM images of Polystyrene/Acorn cupule composites, **g** PS + 1% AC; **h** PS + 5% AC; **i** PS + 10% AC

In (Fig. 2a) Polystyrene has a compact and smooth shape, for the reinforcing particles, we notice that the acorn cupule (Fig. 2c) has a porous structure with asymmetrical fibers, however, the acorn pericarp (Fig. 2b) has a composition with an elongated morphology as well as a more intact and firm structure.

After the insertion of AP and AC with different percentages (1%; 5% and 10%), we observed remarkable changes in the morphology, from a compact polystyrene to a material that adopts a laminated morphology according to the percentage of reinforcement added for the PS/AP composites (Fig. 3), also for the PS/AC composites we observed that the particles of Acorn cupule inserted in the polystyrene matrix changed the morphology of composite to a porous structure (Fig. 4).

4.2 Effect on Crystallography: X-ray

To investigate the effect of incorporating two reinforcements; AP and AC, on the PS structure, X-ray diffraction measurements were performed. Several investigations have shown that practically all crystalline polymers contain amorphous regions. The presence of amorphous regions results in characteristic amorphous halos in the diffraction pattern. Figure 5 shows the X-ray diffraction pattern of the polyester sample. From the figure, we can see that the X-ray diffraction pattern of the PS is characterized by a halo extending from $2\theta = 8°–25°$, with a maximum peak at 17°. The outline of the halo indicates that the polystyrene is a partly crystalline polymer as reported by [5]. There are two types of polystyrenes: atactic polystyrene and syndiotactic polystyrene. One is crystalline, the other is very amorphous. As seen from (Table 2) that synthesis PS is very crystalline with a crystallinity of 47% which means that we have a Syndiotactic PS. Syndiotactic polystyrene is highly ordered, with the phenyl groups alternating on either side of the chain. This type of polystyrene has attracted the attention of several studies thanks to its mechanical properties and its very interesting chemical resistance [6, 7].

Figure 6 evidence the X-ray diffractogram of AP and AC reinforcements, which presents two peaks well defined. These peaks are situated at $2\theta = 15°$ and $2\theta = 21°$,

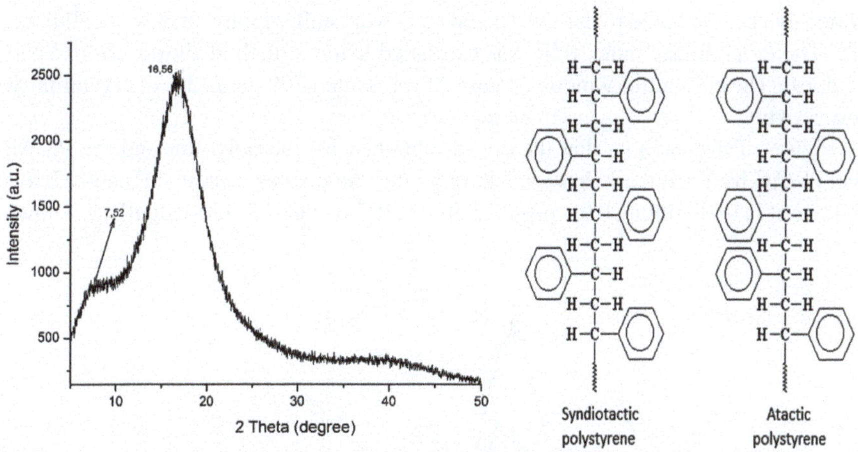

Fig. 5 X-ray diffractogram of PS (to the left) and structure of PS (to the right)

Table 2 Crystallinity of PS, PS/AP composites and PS/AC composites, R: AP or AC

Content	Xc%	
	AP	AC
PS	47,57	47,57
PS + 1wt%R	46,92	47,54
PS + 5wt%R	45,73	45,55
PS + 10wt%R	43,57	43,34
R	20,32	14,63

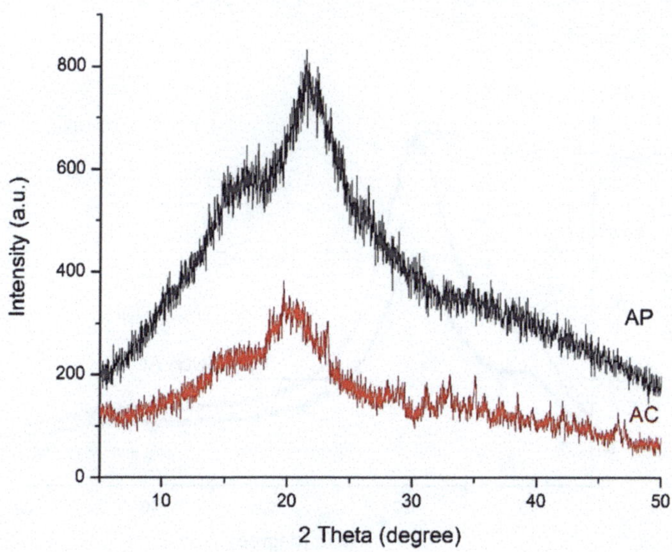

Fig. 6 X-ray diffractogram of AP and AC

which can be attributed to the (2 0 0) and (1 1 0) crystallographic planes of cellulose, [8]. The crystallinity index (CI) was estimated using a method similar to Mulinari et al. [9]. The crystallinity index AP and AC presented 20% and 15% of crystallinity, respectively.

Figures 7 and 8 show that the curves obtained for the polyester and the PS/AP and PS/AC have similar behavior. Although the composites contain AP and AC, it is not possible to distinguish the peaks related to these materials individually, as shown

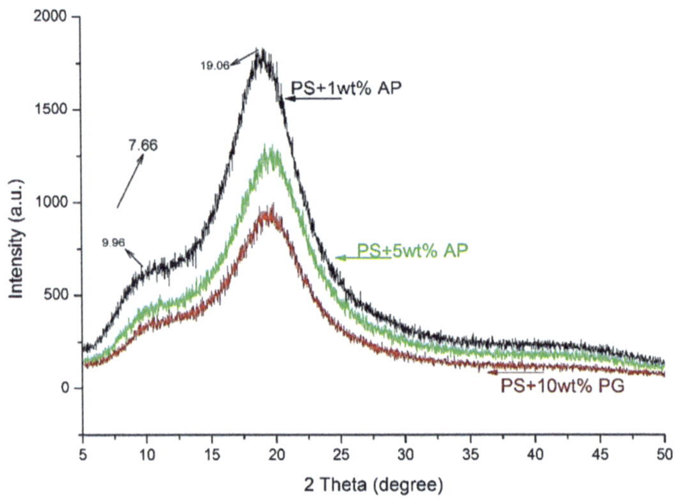

Fig. 7 X-ray diffractogram of PS/AP composites

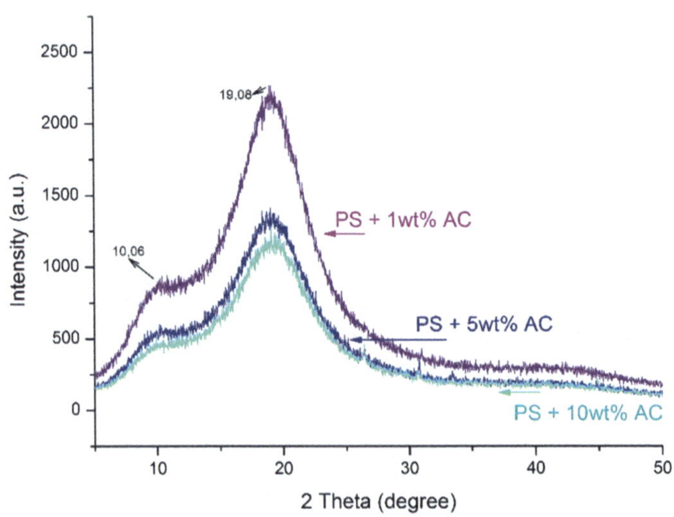

Fig. 8 X-ray diffractogram of PS/AC composites

in (Figs. 7 and 8), this phenomenon was observed by [10]. We note a deviation of the peaks towards the upper angles at $2\theta = 10°$ and $2\theta = 19°$, an indication of the integration of the cellulose within the macromolecule PS with a reduction in intensity, which produce a reduction in crystallinity of composites as be seen in (Table 2).

4.3 Effect on the Structure of PS: FTIR

Figures 9 and 10 show infrared spectra of PS, AP, AC, and both composites PS/AP and PS/AC. We can confirm that we have a syndiotactic PS from the FTIR spectrum

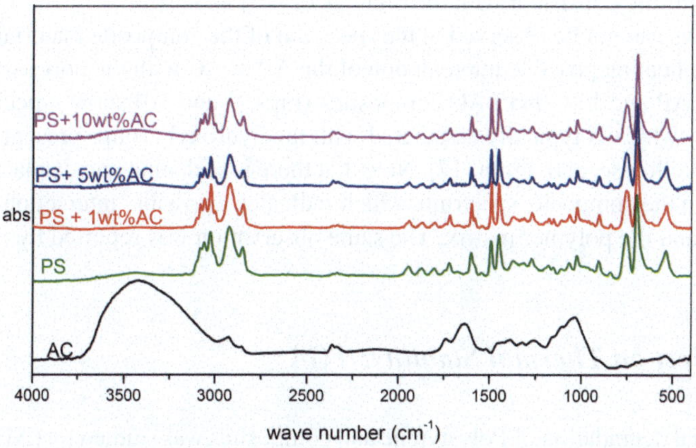

Fig. 9 FTIR spectra of PS, AC, and PS/AC composites

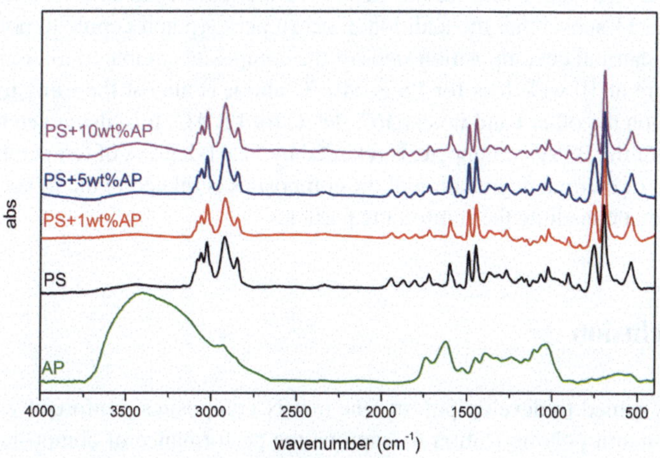

Fig. 10 FTIR spectra of PS, AP, and PS/AP composites

(Figs. 9 and 10) [11], the bands at 1602 and 1583 cm-1, attributed to the ring C–C stretching and H-C–C bending as reported by [12, 13]. In addition, the peaks between 1000 and 600 cm^{-1} are assigned to a ring H–C–C bending, the C–H out-of-plane bending, and a ring C–C torsion [11]. It can be seen from the two Figures that AP and AC had a similar spectrum IR. A very large and strong band from 3500 to 3400 cm^{-1}, is attributed to the elongation vibrations of the hydroxyl (-OH) groups of cellulose and hemicellulose [14]. The bands at 2800 cm^{-1} and around 1600 cm^{-1}, attributed to the stretching of symmetrical CH groups and stretching of unconjugated CO groups present in polysaccharides and xylans [8]. The weak bands at 1262 cm^{-1} and 1161 cm^{-1} correspond to the (C-O) elongation of lignin and the antisymmetric (C–O–C) elongation of cellulose and hemicellulose, respectively. The doubling of the peak at 1032 cm^{-1} is attributed to (C–O) elongation of cellulose and hemicellulose. These results are consistent with the findings of [15] and [16].

All peaks can not be observed in the spectrum of the composite material (Figs. 9 and 10), indicating possible intercalation of the AP or AC with the polystyrene. The PS/10wt%AP and PS/10wt%AC composites (Figs. 9 and 10) show a peak around 3450 cm^{-1} which is typically associated with the hydroxyl groups present in cellulose, hemicellulose, and lignin [17]. Note that there is widening and displacement of this peak in the composite spectrum, which indicates a possible interaction between the fibers and the polymer matrix. The same observation was reported by [17].

4.4 Effect on Thermal Stability: TGA

The thermal degradation of Polystyrene and composites was studied by (ATG) under an air atmosphere up to 600 °C at a heating rate of 10 °C/min. Two stages of thermal decomposition are found in the pure polystyrene sample. The first step is the removal of moisture, followed by the second step of the thermal degradation of polystyrene.

The Fig. 11 shows that the addition of acorn pericarp and cupule to polystyrene delays the thermal decomposition steps of the composite compared to virgin ps. the Temperature at 10 wt% loss for PS is 340 °C and it is almost the same for PS/AP composite on the other hand it is at 367. 94 °C for PS/AC. It is also noted that at 50 wt% loss for the PS/AC composite is retarded by 12.41 degrees of temperature. This retardation of thermal degradation of the composites is related to the lignocellulosic materials that constitute the reinforcing particles.

5 Conclusion

This study aimed to develop polystyrene matrix composites reinforced with acorn waste by in situ polymerization to improve the performance of composites and in particular reduce the risk of plastic on the environment. This bio-composite has been analyzed by various tools and techniques of characterization which present

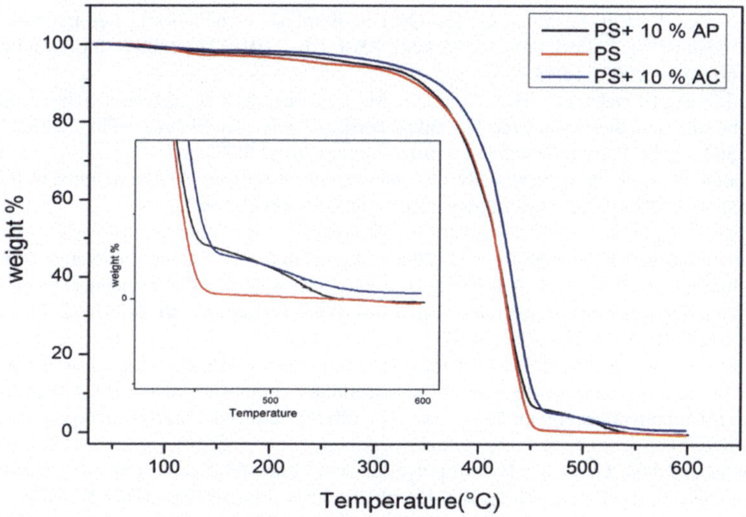

Fig. 11 TGA curves of polystyrene and PS/(AP, AC) composites

good interactions matrix/reinforcement. EDX analysis indicates the integration of cellulose into the PS macromolecule with a reduction in intensity, which produces a reduction in the crystallinity of the composites by a deviation of the peaks to angles greater than $2\theta = 10°$ and $2\theta = 19°$. The recorded FTIR spectrum indicates possible intercalation of AP or AC with polystyrene shown by the peak around 3450 cm-1 which is typically associated with hydroxyl groups present in cellulose, hemicellulose, and lignin, and a broadening and shift of this peak are observed. Also, the thermal degradation of polystyrene and composites was studied, the addiction of acorn pericarp and cupule to polystyrene delays the thermal decomposition steps of the composite compared to virgin PS, this delay in the thermal degradation of composites is related to the lignocellulosic materials that constitute the reinforcing particles. SEM micrographs show that the acorn waste particles inserted into the polystyrene matrix changed the morphology of the composite into a porous structure for PS/AC and a laminated morphology for PS/AP.

References

1. Maghnia, F.Z., et al.: Impact of cork oak management on the ectomycorrhizal fungal diversity associated with Quercus suber in the Mâamora forest (Morocco). Comptes Rendus Biologies **340**(5), 298–305 (2017). https://doi.org/10.1016/j.crvi.2017.04.001
2. Koenig, W.D., Haydock, J.: Oaks, acorns, and the geographical ecology of acorn woodpeckers. J. Biogeogr. **26**(1), 159–165 (1999). https://doi.org/10.1046/j.1365-2699.1999.00256.x
3. Li, S., Wang, C., Chu, F., Xia, J., Xu, Y.: Effects of compatibilizers on composites of acorn shell powder and low density polyethylene. BioResources **8**(4), 5817–5825 (2013). https://doi.org/10.15376/biores.8.4.5817-5825

4. Li, S., Xu, S., Liu, S., Yang, C., Lu, Q.: Fast pyrolysis of biomass in free-fall reactor for hydrogen-rich gas. Fuel Process. Technol. **85**(8–10), 1201–1211 (2004). https://doi.org/10.1016/j.fuproc.2003.11.043
5. Nouh, S.A., El-Nabarawy, H.A., Abutalib, M.M., Bahareth, R.A.: Effect of X-ray irradiation on the structure, thermal and mechanical properties of polyester. EPJ Appl. Phys. **62**(3), 30201-p1–30201-p5 (2013). https://doi.org/10.1051/epjap/2013130050
6. Chatani, Y. et al.: Structural study of syndiotactic polystyrene: 1. Polymorphism. Polymer **33**(3), 488–492 (1992). https://doi.org/10.1016/0032-3861(92)90723-A
7. Baugh, L.S., Schulz, D.N.: Discovery of syndiotactic polystyrene: its synthesis and impact. Macromolecules **53**(10), 3627–3631 (2020). https://doi.org/10.1021/acs.macromol.0c00350
8. Mulinari, D.R., Baptista, C.A.R.P., Souza, J.V.C., Voorwald, H.J.C.: Mechanical properties of coconut fibers reinforced polyester composites. Procedia Eng. **10**, 2074–2079 (2011). https://doi.org/10.1016/j.proeng.2011.04.343
9. Mulinari, D.R., Voorwald, H.J.C., Cioffi, M.O.H., da Silva, M.L.C.P., Luz, S.M.: Preparation and properties of HDPE/sugarcane bagasse cellulose composites obtained for thermokinetic mixer. Carbohydrate Polym. **75**(2), 317–321 (2009). https://doi.org/10.1016/j.carbpol.2008.07.028
10. García del Pino, G., et al.: Hybrid polyester composites reinforced with curauá fibres and nanoclays. Fibers Polym. **21**(2), 399–406 (2020). https://doi.org/10.1007/s12221-020-9506-7
11. Albunia, A.R., Musto, P., Guerra, G.: FTIR spectra of pure helical crystalline phases of syndiotactic polystyrene. Polymer **47**(1), 234–242 (2006). https://doi.org/10.1016/j.polymer.2005.10.135
12. Musto, P., Tavone, S., Guerra, G., De Rosa, C.: Evaluation by Fourier transform infrared spectroscopy of the different crystalline forms in syndiotactic polystyrene samples. J. Polym. Sci. Part B Polym. Phys. **35**(7), 1055–1066 (1997). https://doi.org/10.1002/(SICI)1099-0488(199705)35:7<1055::AID-POLB5>3.0.CO;2-V
13. Guerra, G., Musto, P., Karasz, F., MacKnight, W.: Fourier transform infrared spectroscopy of the polymorphic forms of syndiotactic polystyrene. Die Makromolekulare Chemie **191**(9), 2111–2119 (1990). https://doi.org/10.1002/macp.1990.021910914
14. Yang, H., Yan, R., Chen, H., Lee, D.H., Zheng, C.: Characteristics of hemicellulose, cellulose and lignin pyrolysis. Fuel **86**, 1781–1788 (2007). https://doi.org/10.1016/j.fuel.2006.12.013
15. Zhou, F., Cheng, G., Jiang, B.: Effect of silane treatment on microstructure of sisal fibers. Appl. Surface Sci. **292**, 806–812 (2014). https://doi.org/10.1016/j.apsusc.2013.12.054
16. Lu, N., Oza, S.: Thermal stability and thermo-mechanical properties of hemp-high density polyethylene composites: effect of two different chemical modifications. Compos. Part B Eng. **44**(1), 484–490 (2013). https://doi.org/10.1016/j.compositesb.2012.03.024
17. Tomczak, F., Sydenstricker, T.H.D., Satyanarayana, K.G.: Studies on lignocellulosic fibers of Brazil. Part II: Morphology and properties of Brazilian coconut fibers. Compos. Part A Appl. Sci. Manuf. **38**(7), 1710–1721 (2007). https://doi.org/10.1016/j.compositesa.2007.02.004

Design and Analysis of a Smart Anti-lock Braking System

Salahaddin M. Sahboun and Abdulrahman A. A. Emhemed

Abstract The paper is to model and simulate the performance of an Anti-Lock braking system (ABS). In this paper it is also aimed to analyze and test various tyre models and to show the effect of slip, camber, braking, traction, etc. and to produce graphical plots for lateral acceleration, longitudinal traction and braking force. It is also reduce self-aligning moment for both time response and slip angle/ratio response and also for various load conditions. The simulation of the system was carried out by applying these quarter loads and the behavior of the tire was studied to create smart ABS system. The variation of the parameters such as the aligning moment and the slip angle were studied and the obtained results can be considered quite good.

Keywords Smart ABS system · Quarter vehicle · Slip angle

1 Introduction

When brakes are applied, a skidding wheel, where the tire contact patch is sliding relative to the road, has less traction than a non-skidding wheel. concept of Anti-lock braking system is to prevent the wheels from skidding while braking which helps in stopping the vehicle faster and also to improve steering while braking [1–3]. There are four main components to an ABS system, Speed sensors, Pump, Valves and Controller. The speed sensors detect the speed of the wheels when braking is applied and inform the Anti lock braking system when the wheel is about to lock. The speed sensors are located at each wheel. Various sensors can be used to provide this information and in some cases the differential acts as a speed sensor and provides the information to the Anti lock braking system. Each brake controlled by the ABS has valves fitted on the brake line [4, 5]. The valve helps direct the pressure from the master cylinder for braking. The valves have commonly three positions:

S. M. Sahboun · A. A. A. Emhemed (✉)
Faculty of Technical Engineering, Bright Star University, El-Brega, Libya
e-mail: abdo_83f@yahoo.com

S. El Himer et al. (eds.), *Artificial Intelligence of Things for Smart Green Energy Management*, Studies in Systems, Decision and Control 446,
https://doi.org/10.1007/978-3-031-04851-7_14

- In the first position the valve is completely open and the pressure is transferred from the master cylinder to the brake directly.
- In the second position the valve closes the line thus isolating the pressure existing in the brake and preventing further rise in the pressure should the driver press the brake harder.
- In the third position the valve opens up slightly to release some pressure from the brake depending on the requirement.

The function of the pump is to increase the pressure in the brake line since the valve releases some pressure in its third position. In other words the pump is used to maintain the pressure constant in the brake line for effective braking. The controller is usually the ECU which is used to control the speed sensors and also to regulate the pressure flow through the valves depending on the requirement [6, 7].

2 Types of Anti-lock Braking Systems

Anti-Lock braking systems are classified in many different ways but the most general form of classification is based on the number of channels and the number of speed sensors. In this type of ABS there is a speed sensor on each wheel and every wheel has its own valve to regulate the braking pressure flowing to it. Hence the ECU can monitor each wheel individually for optimum performance and achieve maximum braking force [8].

2.1 *Three-Channel, Three-Sensor ABS*

In this type of ABS there is a speed sensor and a individual valve for each of the front wheel and a common speed sensor and valve for both the rear wheels located in the rear axle. This type of system is usually used for pick up trucks with four wheel ABS. The system is effective over all but there is always a chance of one rear wheel getting locked during a stop reducing brake effectiveness since for the ABS to get activated in the rear both wheels have to lock up simultaneously [9].

2.2 *One-Channel, One-Sensor ABS*

This system is commonly found in vehicles with rear wheel ABS. It is provided with one speed sensor and one valve to both the rear wheels located in the rear axle. The working of this system is similar to that of the rear sensor of a three-channel, three-sensor ABS [10, 11].

3 Working of an ABS System

Though there are many variations and control algorithms used for ABS systems the following is an example of how a simple ABS system performs some of the standard operations. The controller with the help of speed sensors senses the abnormal decelerations in the wheels at all times. Since right before a wheel locks the slip suddenly increases there is usually a sudden deceleration in the wheel velocity and since the wheel in such conditions would stop much quicker than the car.

The ABS controller picks up the signal from the speed sensors of this rapid deceleration so it releases the valve and reduces the brake pressure until it sees acceleration. All this process is carried out quickly before the speed of the tire reduces significantly. The result is that the tire slows down at the same rate as the car, with the brakes keeping the tires very near the point at which they will start to lock up. Hence the performance of braking is enhanced [12].

4 Modelling of Quarter Vehicle

The quarter vehicle to be tested for the ABS algorithm was modelled using ADAMS A view software. The diagram of the model to be developed using the ADAMS software is shown in Fig. 1. First, the model is to be created without gravity acting on it. 'MKS' unit system was used for construction of the model [13]. A road over which the model is to run was imported from the ACar shared files. The tire to be tested for the algorithm was imported from the ACar shared files. Two carrier parts was included to the model named car_1 and car_2. The car_1 was connected to the tire by means of a revolute joint placed in the tire 'cm' in the direction of motion of the vehicle. The part named car_2 is connected to the part car_1 by means of a revolute joint placed in the 'cm' of the car_1. A part named 'slider' is connected to the car_2 by means of a cylindrical joint is placed directly below the part car_2. The cylindrical joint is placed between the parts car_2 and the slider. The slider is also

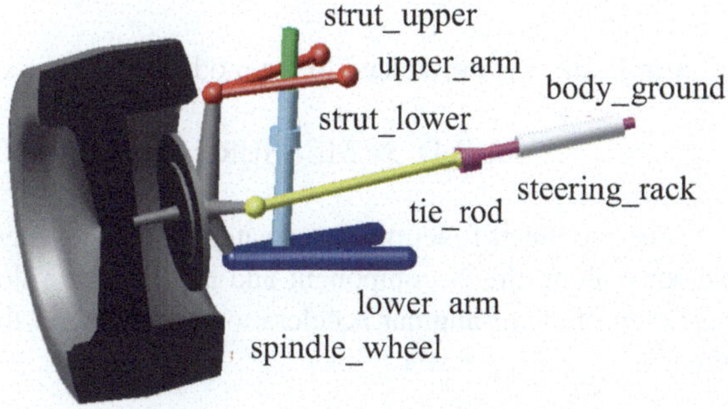

Fig. 1 Modelling of a quarter vehicle

connected to the ground by means of a translational joint placed in the 'cm' of the slider.

The revolute joint between the parts car_1 and the tire is provided for giving the tire a rotational motion. A camber angle during the simulation of the test process will be set by giving a motion input to the revolute joint between the two parts car_1 and car_2. When a motion input is given to the cylindrical joint between the car_2 and the slider during the tire testing process, the required slip angle will be set. The forward motion velocity for the whole system will be set when a motion input is given to the translational joint between the slider part and the ground.

Motions were applied to all the joints. The revolute joint between the parts car_1 and the tire was given a rotational motion of initial functional '0', as for the revolute joint between the parts car_1 and car_2. The only cylindrical joint in the model was too applied with a rotational motion of initial functional value '0'. A translational motion of 27.77 m/s was given to the translational joint between the slider and the ground which accounts for the linear velocity of the vehicle. Two degrees of freedom was achieved at the end of modelling the complete system. The two degrees of freedom were for the slip and camber. A point force was applied on the part car_1 and was located on the tire 'cm'. This point force represents the quarter load of the vehicle.

The complete quarter model consisted of: 2 Gruebler Count (approximate degrees of freedom); 4 Moving Parts (not including ground); 1 Cylindrical Joint; 2 Revolute Joints; 1 Translational Joint; 3 Motions; 2 Degrees of Freedom.

5 Smart Control Algorithm

Slip Ratio: Slip ratio is defined as means of calculating and expressing the locking status of a wheel and is vital to the effectiveness of any anti-lock braking system [14–16]. Slip Ratio is calculated as ((Linear Velocity − Angular Velocity) * Effective Radius of the wheel)/Linear Velocity.

$$\text{SlipRatio} = \frac{(V_x - \omega z) \times R}{V_x} \tag{1}$$

Slip Ratio was introduced to the modelled quarter vehicle in ADAMS using:

BUILD ≫ MEASURE ≫ FUNCTION ≫ NEW

A user-defined function was created within the function builder using the linear velocity along the X-component and the angular velocity along the Z-component (or) magnitude of angular acceleration according to the local coordinate system of the tire.

5.1 SMART ABS Algorithm

First an 'IF' statement was in included to apply the brake pressure on the tire. This was done by:

BUILD >> SYSTEM >> ELEMENTS >> DIFFERENTIAL >> EQUATION >> NEW

A differential function with the name 'DIFF 1' and an initial condition '0' was created and the following 'IF' condition was inserted in the function builder as:

IF [Vx: 0, 0, STEP (Time, 0.5, 0, 0.6, 2500)]

- *Explanation of 'IF' statement:*
- *IF [Exp1: Exp2, Exp3, Exp4]*
- *Exp1 less than 0, it returns Exp2*
- *Exp1 less than 0, it returns Exp3*
- *Exp1 greater than 0, it returns Exp4.*

The value returned by IF [Vx: 0, 0, STEP (Time, 0.5, 0, 0.6, 2500)] is the step function because the linear velocity is considered to be constant at 27.77 m/s which is greater than 0 and hence above statement returns the value of expression 4 which is the STEP function in this case. In the above IF statement, the value 2500 represents the driver input i.e. the brake pressure.

Another differential function with the name 'DIFF 2' and initial condition '0' was created was used to cycle the ABS pressure in accordance with the slip ratio. It is expressed as:

- IF [Time −1.22: 0, 0, IF (Slip −0: 23: ABS On, ABS On, STEP (Slip, 0.23, ABS On, 0.25, ABS off)]
- ABS On = Differential Value (DIFF 1)
- ABS OFF = Differential Value (DIFF 1) * −1.

The cycling pressures of the ABS system are held in ABS On and ABS Off system variables. The above IF function describes when the Time −1.22 is greater than 0, then the IF function within it is executed. For the slip to increase to a certain extent a time delay of 1.22 s is given. When the slip ratio ≤0.23, then the ABS On will be executed. If the slip ratio >0.23, then the STEP function will be executed. In this STEP function, if the value of slip ratio is 0.23, then the ABS On pressure will be activated. If the slip ratio is 0.25, ABS Off will be activated. The above statement means that the ABS pressure is activated and deactivated for the slip ratios of 0.23 and 0.25 respectively.

5.2 *Brake Torque*

The braking torque is expressed as:

$$\text{BrakeTorque} = \text{R}\mu\text{P}(\pi\text{r}^2)\text{n} \tag{2}$$

where

P—Brake Pressure (N/mm^2)
A—Brake Piston Area (mm^2)
R—Radius Center of Pad (mm)
n—no. of friction surfaces
μ—Co-efficient of friction.

Brake Torque was applied to the tire by creating a Sforce on the wheel 'cm'. The function of the S force is represented as:

$$\text{FUNCTION} = \text{n} * \mu * \text{A} * \text{R} * \text{Absolute Pressure}$$

where, Absolute Pressure is a variable given by the summation of the algebraic values of 'DIFF1' and 'DIFF2'.

6 Results and Discussion

From the Fig. 2 it is seen that when the time reaches 0.6 s the first differential equation set the driver input to 2500, because of this the slip fall below 0.23 so as per the second differential equation since the value of the slip is below 0.23 the IF statement activates the expression ABS On. Due to which the angular velocity has found to decrease to certain extent. But the actual system should have a the slip curve in Fig. 3 should be oscillating between 0.23 and 0.25 value of the slip ratio because the control algorithm is given in such away that the ABS On pressure is activated when the slip reaches the value of 0.23 and also when it goes below 0.23 so that the slip ration can be increased so as to activate the ABS Off condition. The Model was able to execute the single loop i.e. The differential equations were executed through which the ABS On pressure was applied to the brake torque in the wheel and plotted against the time and slip angle. Figure 4 shows the angular velocity in. respect to time. Figures 5 and 6 presented the aligning moment verses time and slip angle respectively. Figure 7 shows the dynamics of slip angle in respect to time.

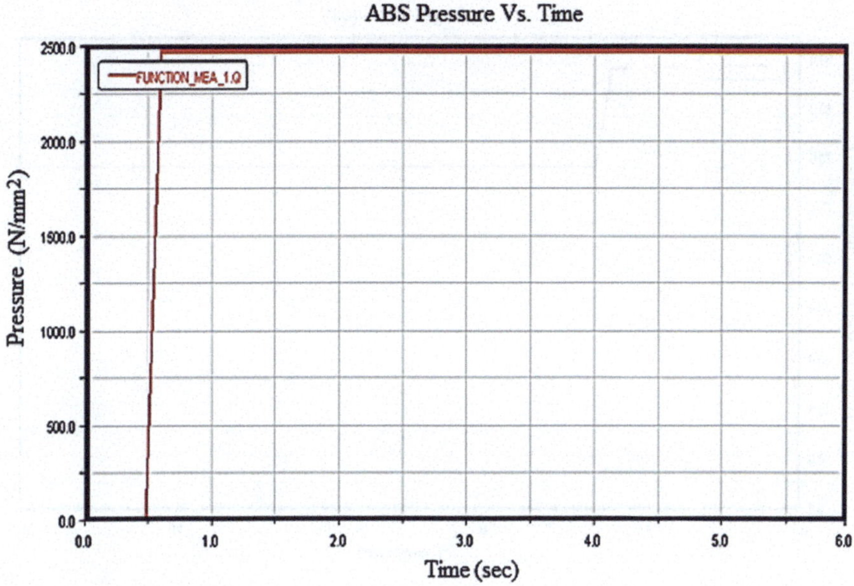

Fig. 2 ABS pressure versus time

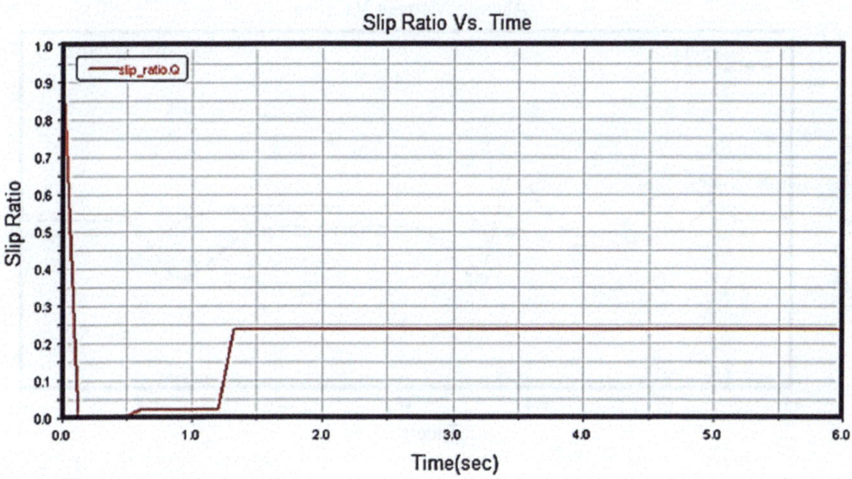

Fig. 3 Slip ratio versus time

7 Conclusion

This paper deals with the Simulation of a smart ABS algorithm using ADAMS AView software. The main objective of the project was to study the behavior of the vehicle system which included the tires, suspension system geometry when the ABS

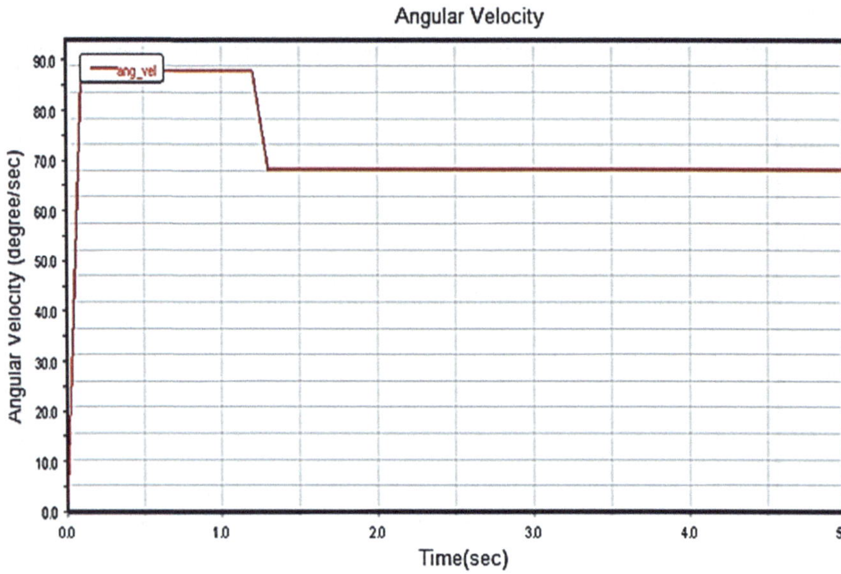

Fig. 4 Angular velocity versus time

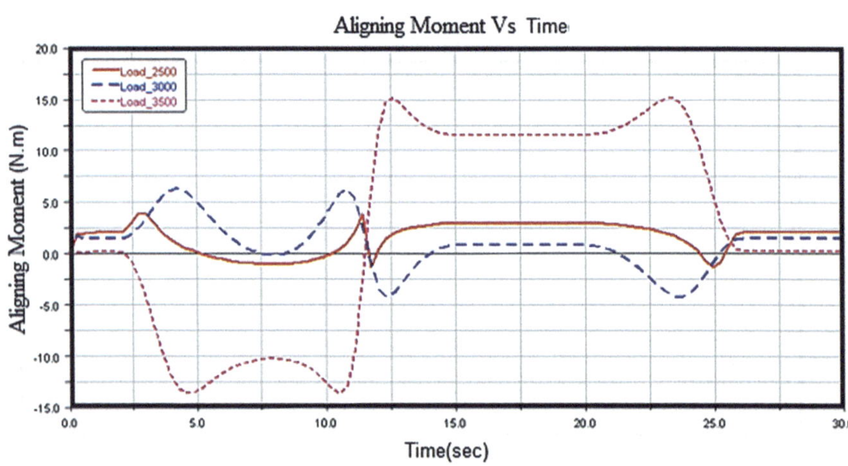

Fig. 5 Aligning moment versus time

algorithm was included in the braking system. First, the quarter model of the system whose behavior is to be studied was modeled using ADAMS AView software. The control algorithms for ABS to be included into the braking system were constructed by building various measures, functions and system variables. The simulation was carried out after the control algorithms were in included into the system. Different parameters of the tire such as the slip ratio, angular velocity were studied. The

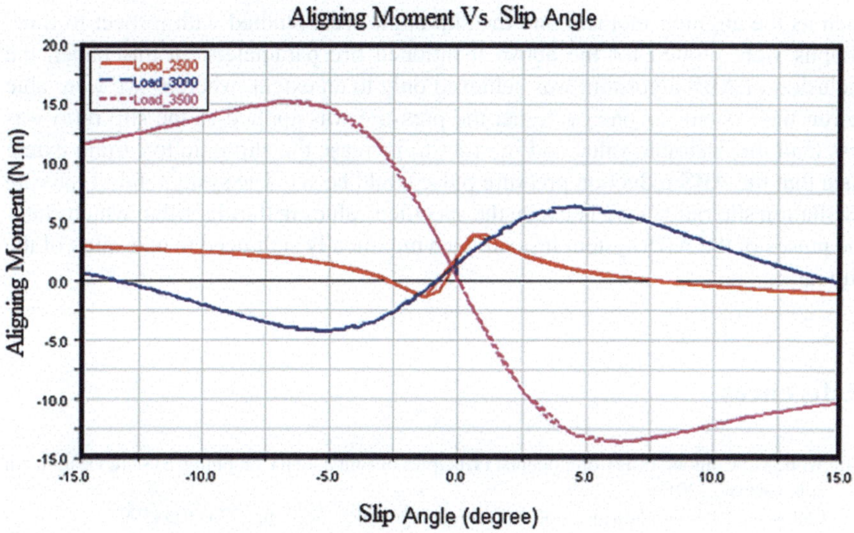

Fig. 6 Aligning moment versus slip angle

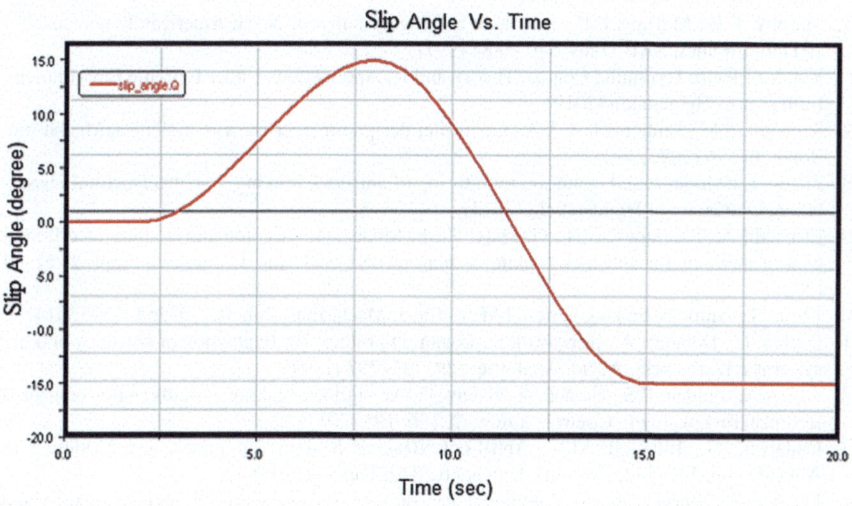

Fig. 7 Slip angle versus time

variation of the parameters over a period was plotted in graphs. The plotted graphs gave us the information about the behavior of the system when the smart ABS was included. The tyre used in the modeled system was tested with different loads. These represented the Vehicle's quarter loads. These loads induced a change in the behavior of the tires smartly. The simulation of the system was carried out by applying these quarter loads and the behavior of the tire was studied. The variation of the parameters such as the aligning moment and the slip angle were studied with respect to time. Graphs were plotted for the above mentioned tire parameters. In this paper, the inclusion of ABS algorithm was achieved only to an extent, whereby we were able to run the system for one cycle i.e. the pressure was applied as the slip ratio was less than the desirable value and in order to increase the slip ratio to certain extent such that the ABS reduction pressure pulse could be set. The system didn't have an oscillation slip ratio curve between the specified values in the algorithm which states the pulses of the ABS system instead it asymptotically stabilized in one value of the slip ratio.

References

1. Wolf, M.: Computers as Components: Principles of Embedded Computing System Design, 4th edn. Elsevier (2016)
2. Gillespie, T.D.: Fundamentals of the Vehicle Dynamics. SAE, pp. 275–303 (1992)
3. Braess, H.-H.: Steering System, Steering System Properties. VDI-Berichte Nr. 1632. SAE Papers, pp. 13–56 (2001)
4. Sandhu, F., Selamat, H., Sam, Y.M.: Antilock braking system using dynamic speed estimation. Jurnal Teknologi **72**, 45–50 (2015)
5. Murphy, B.R., Murhpy, B.R.: Anti-lock braking systems for North American buses, coaches and motorhomes. SAE Trans. 570–588 (1997)
6. Yin, Y.: Electro Hydraulic Control Theory and Its Applications Under Extreme Environment. Butterworth-Heinemann (2019)
7. Sahboun, S.M., Emhemed, A.A.A.: Controller design for steer-by-wire system. J. Mechatron. Robot. **6**, 1–6 (2022)
8. Zheng, L.: Model-based condition monitoring of anti-lock braking systems. Doctoral Dissertation, University of Huddersfield (2014)
9. El-Fatah, A., Sharkawy, A.N., Ghazaly, N.M., Moaaz, A.: A comparative study of different control methods for anti-lock braking system (ABS). SVU Int. J. Eng. Sci. Appl. **2**, 27–34 (2021)
10. Jitesh, S.: Antilock braking system (ABS). Int. J. Mech. Eng. Rob. Res. **3**, 253–159 (2014)
11. Burton, D., Delaney, A., Newstead, S., Logan, D., Fildes, B.: Evaluation of anti-lock braking systems effectiveness. Accid. Anal. Prev. **29**, 745–757 (1997)
12. Aly, A.A., Zeidan, E.S., Hamed, A., Salem, F.: An antilock-braking systems (ABS) control: a technical review. Intell. Control. Autom. **2**, 186–195 (2011)
13. Ozdalyan, B., Blundell, M.V.: Anti-Lock Braking System Simulation and Modeling in ADAMS, pp. 140–144. Coventry University, SAE Papers (1998)
14. Dhivya, P., Murugesan, A.: Intelligent car braking system with collision avoidance and ABS. In: IJCA Proceedings on National Conference on Information and Communication Technologies NCICT, pp. 16–20 (2015)
15. Tomas, B., Brooks, P., Barton, D.: A new analytical model of the brake pad for improved calculation of the centre of pressure and friction coefficient in a multi-piston disc brake. Int. J. Eng. Technol. **7**, 54–63 (2018)
16. Infantini, M.B., Perondi, E., Ferreira, N.F.: Development of an Anti-Lock Braking System Model (No. 2005-01-4032). SAE Technical Paper (2005)

The manufacturer's authorised representative in the EU is Springer Nature Customer Service Centre GmbH, Europaplatz 3, 69115 Heidelberg, Germany. If you have any concerns regarding our products, please contact ProductSafety@springernature.com

Printed and bound by CPI Group (UK) Ltd, Croydon, CR0 4YY
22/09/2026
02221055-0006